Developments
and Changes in
Science Based
TECHNOLOGIES

Developments and Changes in Science Based TECHNOLOGIES

P. MATHUR / K. MATHUR / S. MATHUR

PARTRIDGE

A Penguin Random House Company

To order additional copies of this book, contact
Partridge India
000 800 10062 62
www.partridgepublishing.com/india
orders.india@partridgepublishing.com

CONTENTS

Chapter-3

Visual-Display Devices: Cathode Ray Tube & Liquid-Crystal based Monitor

Chapter-4

Conventional Photography verses Digital Photography

Chapter-5

Chapter-6

Chapter-7
Magnetic Tapes for Audio and Video Recording 79

Chapter-8
Conventional Pendulum Clock, verses Quartz Clock 90

Preface

It is rightly said-, "Today's Science shall be the Tomorrow's Technology".

But, whereas the 'Science' is eternal, a 'Technology' keeps on changing with time. As a consequence of such changes, it is not very uncommon to see an old technology becoming obsolete and being abandoned completely ('out'), or being replaced ('in'), partly or wholly by a new one. But not all the older technologies become obsolete and do not need to be replaced at least for some more time to come. Thus the duration for which a particular technology survives can differ very considerably, compared to some others.

One important factor, on which the development of a new technology depends, is its need and on the 'system', which it has followed. Till about four decades back, the 'Analog-System' has been the common basis for most of the technologies. Then the 'Digital-System' took over and now, the current technologies make maximum use of digital system.

There has also been a general shift, from many completely, manually operated machines, e.g. a pendulum clock or a wristwatch, and a gramophone machine, which were manually energized by winding of a spring, into an electrically driven and digitally operated processes. With a shift to a digital-electronic command and control system, there has been an increase in the efficiency of most of the technological processes.

A typical example is that of a mechanical sound recording, and a sound reproduction system, which was done by a manually powered machine (gramophone). This was replaced by magnetic tape recording (analog system) and playing, and ultimately by digital, audio recording system (compact disc, CD). Exactly the same thing happened to the old pendulum clock and wristwatch.

Certain other examples of technological advancements, which resulted in change from 'old' into a 'new', are the replacement of cathode-ray tube by liquid-crystal display, in TV and computer monitors. Similar has been the case of an incandescent light bulb, being replaced by 'compact fluorescence lamp' (CFL) and sodium and mercury vapor lamps, and coal based steam loco-engine being replaced by diesel and electric engines.

Human memory is short lived. The new generation may not be aware of a once useful technology getting extinct or being replaced, due to the development of a better and stronger, new technology. Examples of such changes are numerous, but here we have only used selected examples to illustrate such changes.

The idea of compiling a record of such selected changes in technology arose from the fact that my grandchildren (three of them) who are technocrats and management trained personals has frequently discussed this topic with me. I have often shown them machines based on old technologies (pendulum clock and gramophone) at our home and their modern substitutes. This created a thought in them to write the articles (chapters) of this book. These were subsequently revised according to my suggestions and edited by me.

The purpose of the present book is to demonstrate such changes, using selected examples only. I hope more of the younger generation shall learn that the technologies, which they are now using, had their old predecessors.

Chapter-1

Analog and Digital Technologies

1.1 Analog, Analogue, Analogous and Analogy

The word 'analogue' is derived from the Greek word 'ανάλογος' (*analogos*), which means 'proportional'. The Dictionary meaning of the word 'analogous' is, 'similar in function'. The term, 'analog' is the 'American-English' equivalent of 'analogue' in 'British-English'. The term 'analogy' in used to indicate similarity in behavior of two dissimilar things. We have, frequently used these terms interchangeably, where it was felt necessary.

As a very simple example of analogy; let us consider the case of a conventional, mechanical (pendulum) clock. With laps of time during any day, the clock-hands continue to change their positions, on the face of the clock, to indicate the time at any particular instant. Hence there is an analogy between, the 'passage of time' on one hand and the 'movement of clock-hands', on the other hand.

1.2 Digital System

The word 'digital' is also derived from the Latin word '*digitus*', which stands for finger. From very ancient times, fingers have been used for discrete counting of numbers (i.e., 0, 1, 2, 3, 4, 5, etc.). Presently the binary 'digital-system', is the most commonly used method, for computing and in electronics. By digital-technology (or the system), it has been possible to convert any information in the real-world (i.e., any written draft, picture, sound, video, etc.) into a 'digital-format'. Thus produced digital information can be stored, modified, recalled, when required and used, with the help of a digital computer. Typical examples of digital-systems are-

1. Digital drafting and writing,
2. Digital audio-recording and playing it back,
3. Digital video-recording and playing it back,
4. Digital photography and display, and many more.

Anyone, who is learning about digital-system, should very clearly understand the difference between 'digital-display' on one hand, and 'digital-system' on the other. The term digital-display indicates that certain results are being displayed in digits (numerically) and not indicated by the positions of an indicator on the dial of an instrument.

1.3 Morse-Code and Digital System

An interesting similarity can be drawn between the 'Morse-Code', which was once used in the conventional telegraphy, and the modern, digital-technology. In telegraphy, an operator uses only two symbols (i.e. it is a binary based language) of the original Morse-Code, i.e. the dot (•), and the dash (-). It has been possible to express any passage written in English, along with numbers and punctuation marks using Morse-Code. More details of Morse-Code and representation of all the English-Alphabets by Morse-Code has been discussed in, chapter-10.

Just like the Morse-Code, the Digital-system also uses only two digits (i.e., 0 and 1) and hence it also a 'binary-system'. By the digital-technology, not only all the alphabets and numbers (in the form of well punctuated words and sentences), but any other electronic signals, which may be based on sound, color, temperature and light intensity etc., can be expressed, stored and recalled digitally and electronically. In a digital-system, certain languages, other than English, e.g. Hindi, can also be used.

Had electronics been so well developed, during the time, Morse-Code was invented, it could have become as good as today's, digital-technology!

1.4 Digital (System) Technology

A digital system is a data based technology, which uses discrete (discontinuous) values. In contrast to this, analog (non-digital) system uses a continuous range of values to represent any information. Although the digital representations are discrete, they can be used to carry out, either a discrete information, such as numbers, alphabets or any other individual symbol, or 'approximations' of continuous information, such as sound, image, and other measurements of any continuous system, e.g. temperature changes.

During past two-three decades, the digital-system has fast displaced analog-system in many technologies. Thus, the digital-system has now been adopted in many fields, of which 13, more important categories have been listed later.

1.5 Analog Signals

Many mechanical or electrical signals are meant for conveying certain information. An analog signal conveys information, using a particular attribute of a medium. As a typical example, an Aneroid Barometer uses the angular position of its needle on a dial, as a mechanical signal, to convey the information regarding the atmospheric pressure and any change there in. Similarly a mechanical clock also produces analog time signals, by the continuous movement of the time-hands on the clock-face. The position of its hour and minute hands, on the face of the clock-dial, indicates the time at any particular instant of day. Similarly, in case of mercury in a glass thermometer, the position of mercury thread on the celebrated glass tube is used to indicate the temperature.

Change of voltage, current, frequency and electrical charge are examples of electrical analog signals. These signals can be read by specific measuring-meters e.g., a voltmeter and an ammeter. The change of such electrical signals (voltage and current), has been used to represent several other information also. The information conveyed from some physical changes, e.g. the mechanical-motion of a vehicle (e.g. an automobile), sound, light, temperature, and pressure can easily be converted into an analog, electrical signal.

1.6 Transducer, Recording and Storage of Analog Information

A 'transducer' is a device by which, it is possible to convert one type of energy-signal (input) into another form of energy-signal (output). Thus, an automobile speedometer, which is an electronic instrument, measures the rate of mechanical motion of a car. Yet another typical example is that of a microphone (transducer), which converts sound (energy) into an electrical impulse. A loud-speaker is also an example of a transducer, because it reconverts an electrical impulse into sound. A photoelectric cell is a devise which coverts light (energy) into a measurable electrical signal.

Magnetic tapes have proved important tools for recording and storage of analog information. Earlier to the introduction of digital-system these were the most important and largely used mode of recording, storage and recalling of information (data). These were extensively used for audio and video recording, and in analog computers.

1.7 Designing Difficulties for Measuring Analog Signals

Numerous meters have been designed and used to measure analog signals. These analogue circuits are difficult to design and require more skill, when compared to digital-systems. This has been one of the main reasons as to, 'Why digital-systems are now becoming more common and replacing older analogue devices?'

An analogue circuit needs to be designed manually, and the process is generally far less automated compared to a digital system. An analog interface is however always needed, when a digital, electronic-device has to interact with the real world i.e. an operating person. For example, a digital radio-receiver, which interacts with man, has an analogue pre-amplifier (for sound) as the first stage in the receiving chain.

1.8 Conveying Data via Analog Signals

An analog signal shall take a value, proportional to certain attributes, which it represents. At any value of a signal, the information conveyed by it, shall be proportional to the magnitude of that attribute, which it represents. The change in the magnitude of a signal is thus meaningful at each of its level. The signal continuously represents different levels of the phenomenon that occurs during a change. As a typical example, let us presume that an electrical signal represents temperature, which is expressed in terms of voltage. Further, let us presume that one-volt signal represents one degree Celsius. In such a case, a signal of 10 volts shall represent 10° C, while a 20 volt signal shall represent 20° C, and so on. Further, a change of 0.1 volts shall correspond to a change of 0.1° C.

1.9 Electronic Analog-System

A continuously variable signal in any electronic system, is called an 'analog-system'. As an example, if we continuously change the electrical-potential (voltage), applied across a fixed resistance, then the variation of the electric-current (ampere), flowing through that resistance, shall be proportional to the applied potential. In other words we say –

'When there is a change in applied potential, there is an analogous (or corresponding) change in current'. This is the expression of the well known 'Ohm's Law', which is mathematically expressed as—

Potential (Volt) / Current (Ampere) = Resistance (Ohms).

Another method of conveying an analog signal is to use modulation. In this method, some base carrier signal has one of its properties altered. Thus 'amplitude modulation' (AM) in a radio, involves altering the amplitude of a sinusoidal voltage waveform and it is the source information. Similarly 'frequency modulation' (FM), is another way of conveying information, which is based on change of frequency. Yet another technique, called 'phase modulation' is based on changing the phase of the carrier signal, and this is also an analog-system.

In analog recording of sound, the variation in pressure created by sound waves, striking on a vibrating diaphragm, of a microphone, creates a corresponding variation in the current, or the voltage passing through it. A change in the volume (intensity and frequency) of the sound causes a fluctuation of the current or the voltage to increase proportionally, while keeping the same waveform or shape. Hence a microphone is a typical example of a transducer.

1.10 Noise (Disturbances, resulting in Change of Signal)

The way any information, in the form of an electrical signal, is encoded by an analogue circuit, is more susceptible to noise, (i.e. disturbance, which can cause change of signal) compared to that in digital circuits. Small change in a signal can result in a significant change in the information represented by that particular signal, and sometimes it can even cause the information to be lost completely.

Since a digital signal can only take, one of two different values (digits, 0 and 1) any disturbance would have to be nearly one-half the magnitude of the digital signal to cause an error. This is the property of digital circuits, which has been exploited to make the processing of the digital signals more noise resistant. In case of digital electronics, the information is, (in a way) 'quantized' and as long as the signal stays within a range of values, it represents the same information. A digital circuit uses this principle to regenerate the signal at each 'logic-gate', and thus it reduces, or completely eliminated noise.

Mechanical, pneumatic, hydraulic and many other systems also use analogue signals, to which some 'noise' is always associated.

1.11 Inherent Noise

Analogue systems invariably have some noise. These are random disturbances, or variations. Some of these are caused by the random, thermal vibrations of atomic particles, present in a system. All these variations of an analogue signal are significant, and any disturbance, which is equivalent to a change in the original signal, shall appear as a noise. As the signal is repeatedly copied or transmitted over long distances, these random variations become more significant and lead to deformation of the original signal. Other sources of noise may include external electrical signals or poorly designed components in an electrical circuitry. These disturbances are reduced by proper shielding and by using low-noise amplifiers.

1.12 Precision of a Signal

'How precise is a signal?' This depends upon a large number of factors. The noise present in the original signal, and the noise added during processing it, when taken together contributes to the 'precision'. Precision is determined by 'signal-to-noise' ratio. Fundamental physical limits of components in a circutry, determine the resolution of an analogue signal.

In digital electronics, additional precision is obtained by using digits to represent any signal. A practical limit in the number of digits to be used, is determined by the performance of an 'analogue-to-digital converter' (ADC), because digital operations can usually be performed without loss of precision. The ADC takes an analogue signal and changes it into a series of binary numbers. An ADC may also be used in simple digital-display devices e.g., in a digital-thermometer, or a digital light-intensity meter. It can also be used in digital-sound recording and in digital data acquisition.

Similarly, a 'digital-to-analogue converter' (DAC) is used to change a digital signal to an analogue signal, for display. A DAC takes a series of binary numbers and converts these into an analogue signal. It is very common to find a DAC in the gain-control system of an operational amplifier, which in turn may be used to control digital amplifiers and filters.

1.13 Analogue *versus* Digital Electronics

Any information is encoded differently in analogue and digital electronics. Hence the way in which they process an electrical signal is also different. All operations that can be performed on an analogue signal such as amplification, filtering, limiting, etc., can also be duplicated in the digital-system. Every digital circuit is also an analogue circuit, in so much as the behavior of any digital circuit can be explained using the rules of analogue circuits.

Most of the early electronic devices, which were invented and subsequently mass-produced, were based on analogue system. The use of microelectronics, in digital systems has very much reduced the cost of digital-techniques. Hence, many digital methods are now feasible and these are also cost-effective. Typical examples are those, which are used in the field of 'human to a machine communication by voice'.

1.14 Technologies, where Digital Systems are now adopted

Digital-systems are fast replacing analog-systems in many technological fields, of which 13 main categories are listed below—

1. Digital audio.
2. Digital broadband.
3. Digital broadcasting.
4. Digital computing.
5. Digital humanities.
6. Digital media.
7. Digital display connectors.
8. Digital electronics.
9. Digital (or electronic) publishing.
10. Digital movie cameras.
11. Digital photography.
12. Digital radio.
13. Digital software-defined, radio.

1.15 Digital Noise

During handling, or transmitting of any data, certain amount of noise invariably enters into the data-system, which is being transmitted as electrical or electronic signals, or being stored by analog-system. Such noise can be due to several causes. Thus, during the transmission of some data by wireless, and being received by a radio, there may be certain inaccuracies. A particular data may suffer from interferences from numerous other wireless sources, or even pick up some background noise from the vast universe, of which, we are only a small part.

For example, when a microphone is being used in a speech, it can pick up both the intended sound signal (of the speaker) and some background noise (signal) without any discrimination between them. To some extent, such noise is also included, when the audio is converted to digital-signal.

1.16 Noise due to Electrical / Electronic system

The resistance of a connecting wire, which is used to transmit the electric pulses (or the signal), can attenuate these pulses. Such attenuation is caused by the change in capacitance and inductance of the transmitting (system) media. Variation of temperature, during transmission can also increase or reduce these noise-effects.

While the digital transmissions are also subjected to some degradation slight variations do not matter much, because these are ignored, when a digital signal is received. With an analog signal, 'variances' cannot be distinguished from the 'original signal', and hence it provides a kind of distortion of the signal. In a digital signal, similar variances will not matter, because any signal, close enough to a particular value will be interpreted as that value. Care has to be taken to avoid noise and distortion when connecting a digital to an analog system, but more when using analog systems.

Overall it can now be said that, where ever possible, the analog-systems are going to be 'out' and being replaced ('in') by digital systems.

1.17 Use of Electronics and Improvements in Audio / Video Transmission

Most of us are quite aware about, a rather poor quality of audio (radio) and video (TV) transmissions, which we used to receive at our homes, more than two decades back. Major improvement in the quality of these transmissions, has been due to the adoption of digital-technology. To a large extent, any undesirable change of transmission signal quality, which used to arise due to the distance from a transmission center, and those due to weather condition, has been very much reduced by adopting digital technology. At the receiving centers also, the digital machines have now replaced the old machines, which used analog-system. This has been possible due to digital-technology, now extensively being used by most of the manufacturers. In India one very recent change from analog to digital transmission has been done in selected metros.

Maximum use of electronics is now being made for analog as well as digital-technology in their systems. To conclude therefore, in future the digital-system, shall replace analog-system, completely or wherever possible.

Chapter-2

⤝⧓⤞

Incandescent Light Bulb, Compact Fluorescent Lamp, and Sodium and Mercury Vapor Lamps

2.1 Introduction and Historical: Incandescent Light Bulb

Almost till the end of the nineteenth century, whenever it was necessary for a person to work during the dark night hours, some light was generally created by burning a wax-candle in Europe, while a 'Deepak', with some oil and a cotton wick was used to produce light in India. During those days, persons were in a good habit of carrying out most of their work during the day-light hours.

The invention of a workable 'Incandescent Light Bulb' was made by Thomas Elva Edison, in the year 1880. This made a major difference in lighting of man's habitat and work place, which ultimately resulted in a major change in his working and living style also. Edison, patented, his invention of bulb in the same year (1880). After his invention of the incandescent light bulb, Edison is said to have remarked –

"I will make lighting with electric incandescent bulb so cheap, that only the rich persons shall be able to afford the use of candles".

Though the use of electricity for lighting has now become common all over the world, including the developing countries such as India; how cheap is electricity for use to light a bulb and several of its other applications remains debatable?

For making his incandescent light bulb, Edison used carbonized filament, made from bamboo-fibers. It was much later that metallic filament for bulbs were introduced. Bamboo-fiber filament, in Edison's bulb was housed in a sealed, evacuated glass bulb, with its end terminals for connecting to the supply of electricity. Rechargeable storage batteries were used to supply electric current to Edition's bulb.

Common use of incandescent light bulbs only became feasible, when regular production and distribution of electricity was started, in certain countries, on a commercial scale. For over

14

one and a half century, the cheap, effective and easy-to-use, incandescent light bulb proved to be a great technological advancement. It has been the most popular method of bringing light indoors, which has resulted in extended working-hours for a man, after sunset. A typical incandescent light bulb is depicted in **Fig.-1**.

Now, with the starting of twenty-first century, there are indications that the conventional incandescent light bulb will give way (or, has it already?) to more advanced and efficient technologies for producing light using electricity, i.e. *via* the phenomenon known as 'fluorescence' and 'electric-discharge through gases'.

2.2 Earlier Improvements of Incandescent Light Bulb (ILB)

It has been a common observation that most of the metals on heating become incandescent, and start emitting light. Finally, the metal melts at a specific melting temperature or it gets oxidized. The structural bonds between the atoms of a metal in solid state are disrupted due to heat and results in its melting.

To replace the less durable carbon filaments, which were used in the earlier incandescent bulbs, the use of osmium and tantalum metal wire-filaments was then started. These metals have very high melting temperatures. Later on, tungsten metal was found to be the most economical metal for making bulb filaments, because it has an extremely high melting temperature, and yet it is cheaper and also more abundant. In place of evacuating an ILB, filling it with an inert gas (argon) was introduced.

Even before the independence of India, 'Mysore Lamp' and 'Bengal Lamp' companies became pioneer in making ILB. These ILB manufacturing units, which were established just before independence, served the country well, for a long period by manufacturing indigenous products, and saved much needed foreign currency at that time. Much later, Bajaj Bulbs Co. and some other multinational companies e.g. Philips, Seiman, General Electric and Osram also started manufacturing ILB in India.

2.3 Energy Losses in ILB due to Heat

One major limitations of ILB has been the large production of heat-carrying, infrared radiations, along with visible light spectrum. Only 10-15 % of the radiations produced by an incandescent bulbs, fall in the spectrum region, which is visible to human eye. Thus there is a lot of wastage of electric energy, due to the production of unwanted heat radiations.

In contrast to the ILB, fluorescent lamps produce 'cool light', and they do not waste lot of energy to generate heat radiations. They produce radiations mostly in the visible light spectrum region. For this reason, fluorescent lamps are now slowly edging out the old and reliable ILB.

Light bulbs are ranked by the amount of light they produce in certain time. This is measured in terms of a unit called, 'Watt'. Watt is the energy (Jules) used, as expressed in terms of the product of voltage (Volts) and current (Amperes). Thus, for a 50-watt bulb using 220-volt power supply, current flowing through the bulb can be calculated as under-.

(Wattage) 50 = 220 (Voltage) X current (ampere),

Hence the current through the bulb = 50/220, or ~ 0.228 Ampere.

Bulbs with higher wattage have bigger filaments, and they produce more light as well as heat. Thus, in a multi-apartment building in the USA, the use of ILB made a significant contribution towards room heating (~10%), and it has been estimated that more heating of the apartments is now needed, when fluorescence lighting is being used.

2.4 Some Finer Developments ILB Technology

A simple ILB does not, appreciably change its light intensity, except for unwanted, large voltage fluctuations in the power supply line. A need was felt to have a single ILB, which could provide variable intensity of light, when desired. This resulted in development of ILB, which can provide three different intensities of light. Incandescent, three-way bulbs are now made, and these bulbs have two filaments of different wattage, say one of 50 and another of 100 watts. Such filaments are wired to separate power circuits, which can be switched, ('on' or 'off'), as desired using a special, three-way socket. The switch in the three-way socket lets one choose from three different light levels. Thus, a bulb can work on three different (50, 100, and 150) wattages as desired. On the lowest level, the switch is 'on', only to the circuit for the 50-watt filament, while for a medium light level, the switch is 'on' to the circuit for the 100-watt filament. For the brightest level (150-watt) the switch is 'on' to the circuits for both filaments. The bulb now operates at 150 watts. Earlier to the development of such multiple watt bulbs, putting up of a variable voltage transformer in its circuit could vary the light intensity of ILB.

So far, (till the year 2012) it is not yet uncommon to find many ILB in Indian homes, but soon these shall disappear. In case you are interested in collection of antiques, then preserve one or two ILB (in working condition), which shall be your proud possession to be shown (say, in the year 2020) to your children.

There has been a recent, interesting reporting about the exceptional durability of an ILB. One bulb, nearly 110-years old has been reported to be still in working condition at the fire station at Livermore town of California in USA.

2.5 Halogen Lamp

A halogen lamp is also an incandescent lamp with the usual tungsten filament. The filament is enclosed in a high melting, borosilicate glass bulb, having an inert gas (argon) and a small amount of a halogen (iodine or bromine) contained in it. The combination of the halogen and the tungsten filament produces a chemical reaction known as a 'halogen cycle', which increases the lifetime of the filament and simultaneously prevents darkening of the bulb. Any metallic tungsten, evaporating from the filament redeposit back on it, *via,* formation of volatile halides of tungsten. Because of this, a halogen lamp can be operated at a much higher temperature than a normal ILB. The higher operating temperature results in higher, (visible) white-light spectrum of lower wave length. This, in turn, gives it a higher luminous efficacy (10–30 lm/W) to the bulb. Because of their smaller size, halogen lamps can advantageously be used with optical projection systems, e.g. epidiascope, overhead projector and slide-projector. Halogen lamp work efficiently, but needs a cooling fan. In automobile head-lights, small halogen lamps, with metallic reflectors are now used and produce very bright light.

2.6 Emergence of Compact Fluorescent Lamp (CFL)

A common 'tube-light' is a fluorescent lamp, but it requires special lamp fixture and it is not compact. A 'Compact Fluorescent Lamp' is an energy saving device. CFL is basically designed to replace the conventional ILB, which have been in use for over two centuries now. Most of the CFLs can fit into the existing light-fixtures, i.e. those, which were originally designed to be used for an ILB.

A CFL producing the same amount visible light as an ILB uses far less (~20%) power and has a much longer rated life period. All fluorescent lamps contain in them, a small amount of mercury, which complicates the problem of their disposal, when they have fused out. The light spectrum, produced by a CFL is different from that produced by ILB. CFL produces fluorescent light as the result of high-energy radiations (including an electron-beam), striking on the phosphor coat, inside its glass tube.

A 'Phosphor' is a substance, capable of absorbing high-energy radiations and reemitting that energy in the form of visible spectrum. This phenomenon is called 'fluorescence'. A typical example of a phosphor is a fluorescent pigment, synthesized form phthalic anhydride and resorcinol, and has the property of producing yellow-green fluorescence, when it is exposed to visible radiations. Formulations for phosphors, for use in CFL have now been very much improved to produce 'soft-white' light, which is soothing to human eyes.

2.7 Historical Development of CFL

In the late 1890's P. C. Hewitt invented and patented a fluorescent lamp, which was then exclusively used for lighting in photographic studios and in the industries. Patenting of a high-pressure vapor lamp by E. Germer and F. Mayer followed this. Yet another patent on high-pressure vapor lamp was granted to H. Spanner, in the year 1927. Later on (1938-1941), G. Inman created a practically usable CFL, which was marketed by the General Electric Co. (GE, USA). Soon after, this, circular and U-shaped CFLs **(Fig.-2, 3)** were designed, which reduced the length of fluorescent light fixtures. During the 1976's petroleum-oil crisis, E.E. Hammer designed a spiral tube CFL for GE,. The CFL designed in this invention, met very well the goals of electric-power saving. However the GE did not immediately built up a new factory for production of this newly designed lamp. Other manufacturers soon copied this design. It was in the year 1995 that spiral lamps manufactured in China become commercially available. The manufacture and sale of spiral CFLs has, since been increasing all over the world.

In India, the 'tube-light', based on the principle of production of fluorescent light, was introduced in early 1950's, but as stated earlier, it was not 'compact' like a CFL, and required special 'tube-light holder'.

Bajaj Bulb Co., and HPL Electric & power (P) Ltd. Co., in India are currently manufacturing a complete range of CFLs. According to the current manufacturing practice in most of the countries, including India, all the components of CFL, may not be manufactured in a particular country, say India. Economic production of any commercial product, anywhere in the world, now involves import of many components from world over, and assembling the product for a country's requirement and even for export.

2.8 Finer Developments in CFL

A need to develop fluorescent lamps that could fit in the same volume as a comparable incandescent light bulb, required the development of new and high-efficacy phosphors, which could withstand more power per-unit area than the phosphors used in older and larger fluorescent tubes. In any fluorescent lamp, there is a need of 'ballast' (also called choke) for initial discharge through the tube. Old fluorescent tubes used external ballast on the light fixture. Philips Electrical Co., in the year 1980, introduced its CFL, which had inbuilt ballast. Such CFL could directly be fitted into the regular bulb-holder. For producing good visible light spectrum the CFL's now use tri-color phosphors and mercury amalgam. Osram Co. went a step ahead in introducing the first CFL, having built-in electronic ballast.

Initially the CFLs used 'electromagnetic ballast', which has now been replaced by electronic one. This has removed flickering and slow starting of light, which was associated with earlier fluorescent lamps, i.e. the tube-light.

There have been two types of CFLs, i.e. the integrated type and the non-integrated type. Integrated CFLs combine the lamp and electronic ballast in single unit and these are easy to install and replace. Non-integrated CFLs have now been phased out completely. CFL in 3-way dimmable models are also made.

2.8 Construction Details

A CFL has two main parts i.e., a gas filled tube, which is the bulb component and a ballast. Electronic ballast contains a rectifier, filter-capacitor and two switching transistors (connected as high-frequency resonate series), and DC to AC inverter. Around 40 KHz high frequency current is applied to the CFL, and resonate convertor stabilizes the current, which in turn produces stable light, even when there is some fluctuations in the current being supplied. CFLs with magnetic ballast, used to flicker at the start, but this has been completely eliminated with introduction of electronic ballast.

2.9 Power Source for CFL

CFLs are now available for AC as well as for DC electricity power supply. DC-CFLs are particularly used in mobile recreational vehicles and houses, which are not being supplied by AC-power line. CFLs are particularly useful in developing countries like India, where the earlier used kerosene lanterns can be replaced by CFL using rechargeable car-batteries, solar panel (street lamps) or wind energy.

2.10 Energy Consumption and Lifespan of CFL and ILB

On an average, the lifespan of a CFL can be 8-15 times that of an ILB. Thus the CFLs have a rated lifespan of 7,000-15,000 hours, when compared to ILB, which has a lifespan of 800-1,000 hours only.

Besides any manufacturing defect in a lamp, its lifetime also depends on many other factors, e.g. constancy of operating voltage, mechanical shock, frequency of putting the lamp, 'on and off', lamp orientation and ambient operating temperature.

The 'U.S. Energy Star' program has suggested that a fluorescent lamp should be left 'on', when leaving a room for less than 15 minutes. This shall result in increase its life scan. Generally CFLs produce less light in the later period of their lives, when compared to a newly installed one. Towards the end of its life, a CFL can be expected to produce only about 70–80% of the original light for which they were manufactured. However a 20–30% reduction of light intensity is easily compensated for by the human eyes, for most purposes.

2.11 Health issues

As mentioned earlier, the cost effective and battery-powered CFLs are now replacing the kerosene lanterns in village houses and in work places. This is because they are free from kerosene fumes, which can cause chronic lung disorders in the developing countries like India.

According to a study, carried in the year 2008, by the 'European Commission, Scientific Committee on Emerging and Newly Identified Health Risks', CFL could pose a health risk due to some ultraviolet rays emitted by such devices. More research shall be needed to establish whether compact fluorescent lamps constitute any higher risk than incandescent filament lamps. Any individual, who is exposed to the light produced by single-envelope CFL for long periods of time, at a distances <20 cm from the lamp, could lead to ultraviolet exposures approaching the current workplace limits, set to protect workers from skin and retinal damages. However the UV radiation received from double-envelope CFLs largely mitigates other risks from CFLs.

2.12 Alternatives to CFL and ILB for Street Lighting. Sodium Vapor Lamp

In spite of certain limitations, CFL has successfully replaced ILB for in-house lighting. For out-door and street lighting, economizing on energy cost is an important consideration. Hence in large cities and towns, sodium vapor lamps and mercury vapor lamps are now extensively used for out-door lighting.

Unlike CFL, sodium vapor lamp is a gas discharge lamp that uses metallic sodium (vapor) in an excited state to produce light. These lamps can be of low or high pressure type.

Sodium vapor lamps have long been used in science laboratories, as a source of yellow monochromatic light. Its use for in-home lighting is not done, because man is used to day-light, comprising of the visible region of the spectrum (red to violet), which appears as white-light to human eye.

2.13 Construction and Working of Sodium Vapor Lamp

Low-pressure sodium-vapor lamps are made-up of borosilicate-glass, gas-discharge tube, containing metallic sodium and a small amount of neon and argon gases, which helps to start the gas discharge. The discharge tube may be linear or it may be U-shaped. An outer glass vacuum envelope for thermal insulation, improves the efficiency of sodium vapor lamp. A further improvement, in sodium vapor lamp was attained by coating the glass envelope with an infrared reflecting layer composed of indium-tin oxide.

Sodium D-line emission is an intense yellow band produced due to the transition of $3s^1$ electron in sodium atom. Nearly a monochromatic light is produced by sodium vapor lamp, though actually it consists of two dominant spectral lines of very close wave-lengths (589.0 and 589.6nm). The intense yellow sodium light comprises of nearly 90% of the visible light emission for this type of lamp.

On turning on a sodium vapor lamp, initially it emits a dim red/ pink light, which warms up sodium metal. After a few minutes the light turns into bright yellow, due to vaporization of sodium, which is a nearly monochromatic light. Objects, lighted by sodium vapor lamp are not easily distinguished in color, because they appear almost entirely yellow due to the reflection of this narrow bandwidth of yellow light.

Sodium vapor lamps are the most efficient electrically powered source of light, and under photopic lighting conditions, they produce, up to 200 lm/W light intensity, when compared to 10–30 lm/W for ILB. The output light has a wavelength, of peak sensitivity of the human eye. As a result they are now widely used for outdoor lighting.

2.14 Mercury Vapor Lamp

Mercury vapor lamps are, yet another efficient source for street lighting and produce much better spectrum when compared to that of a sodium vapor lamp. These also have a longer lifetime, and produce intense, white light, useful for several special purpose applications. Light intensity of mercury vapor lamps is so intense that these are not suitable for domestic, in-room lighting.

2.15 Invention of mercury vapor Lamp

Charles Wheatstone, John Thomas, and Leo Arons (1835-1860) made initial contributions in discovering the spectrum of mercury vapors and designing of a lamp based on it. A commercially feasible mercury vapor lamp was patented in the year 1901 by P. C. Hewitt. It produced light, having very little red component. The lamp was commonly used for industrial lighting i.e. in factories and photographic studios. Mercury vapor lamp also produces ultraviolet light, which was earlier (1910) used in sterilizing treatment of a city's water supply. In 1930's Osram and General Electric Company made, improved mercury vapor lamps, which were put into use for general, street lighting.

Resistance to electric discharge through mercury vapor in a lamp, decreases with time, hence it requires a ballast to prevent it from taking excessive current, once the discharge has started. The ballast used in mercury vapor lamp is basically similar to those used in CFL. Just like a CFL, mercury vapor lamp also requires a starter, which forms a component of the lamp.

In a mercury vapor lamp, third electrode is mounted near one of the main electrodes and connected through a resistor to the other main electrode. When power is applied, there is sufficient voltage to strike an arc between the starting electrode and the main electrode, adjacent to it. This arc-discharge eventually provides enough mercury in ionized form to strike an arc between the main electrodes.

2.16 Conclusion

It can safely be said that the era of ILB is now over. In many countries, including in India, the production of ILB has been phased out. CFLs are going to replace them completely. Very soon the ILB shall become a museum piece and coming generation shall know about it only through history. Yet the credit to first invent electric-lighting, in the form of ILB, goes to Thomas Alva Edison **(Fig.-4)**

Fig.-1 Typical Incandescent Light Bulb.
Photo-KMU Bentuzer, Article on ILB, Taken from
http://en.wikipedia.org/

Fig,-2 Typical, U-Shaped CFL

From article on CFL, Auther, Armin Kübelbeck,
Taken from http://en.wikipedia.org/

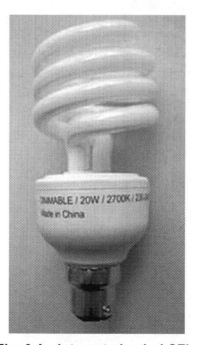

Fig.-3 An Integrated spiral CFL.

Photo—P, Namkel, from article on CFL,
Taken from http://en.wikipedia.org/

Fig,—4, A photograph 0f Thomas Alva Edison
Article on ILB, Photo-M Manske,
Taken from Article on ILB http://en.wikipedia.org/

Chapter-3

❖

Visual-Display Devices: Cathode Ray Tube & Liquid-Crystal based Monitor

3.1 Introduction: Visual-Display Devices

A 'monitor', or a 'visual-display device' is the unit of a personal computer (PC), which displays one's computational work, including figures, pictures, written drafts etc. In case of a television-set (TV) it displays televised shows on the screen. Computational work, displayed on a PC-monitor can be stored and later modified or deleted, as desired. Currently there is a general trend among companies manufacturing TV's or PC's to replace the traditional cathode-ray tube (CRT) based monitor, by liquid-crystal display (LCD) based monitor.

A CRT consists of a large evacuated glass tube, with an arrangement to generate a beam of electrons inside it, which then falls on a phosphor-coated screen. This results in illumination on the outer surface of the phosphor-coated glass-screen of a CRT.

A LCD based monitor comprises of a plastic screen, coated with a liquid crystal material, and suitable electronic circuitry behind it, which is encased in a plastic enclosure.

For TV-sets and computer-monitors, visual devices, based on CRT's were conventionally used, till around the ending of year 2000, when LCD based monitors were introduced, by the manufacturing companies. LCDs are now fast replacing CRT's in most of the applications. LCD's, which comprise of thin, flat electronic visual display devices were already being used in many instrument panels, including aircraft cockpit displays, digital quartz clocks and watches, electronic-calculators, mobile-phones and telephones.

LCD's, besides being lightweight, are less fragile, compact, more portable, less expensive and more dependable, when compared to CRT's. These are also more comfortable to eyes, as compared to CRT-display and do not produce harm-full UV-radiations. They are available in a much wider range of screen sizes compared to

CRT. Since there are no glass component in LCD's they are not fragile.

3.2 Invention of Cathode-Ray Tube (CRT)

Early experimentations on cathode rays were carried by an English—Physicist, Sir, J.J. Thomson. He has also been credited for inventing and carrying out experiments on cathode-ray tube, which resulted in revealing the electronic structure of atom. Sir, Thomson was able to deflect cathode-rays, which consisted of a beam of charged particles i.e., electrons, originating from a cathode, using a magnetic field. This formed the fundamental principle in application of the modern CRT. A German physicist, Ferdinand Braun has also been credited to the invention of an early version of the CRT, called the Braun-Tube, in the year 1897. A cold-cathode diode, which was a modification of Crook's tube, employed a phosphor coated screen in a CRT.

3.3 Description of a Cathode-Ray Tube

A cathode-ray tube (**Fig.-1**), which is used in a TV-set consists of large, cone-shaped, evacuated glass tube, fitted with one or more source of electron-beam. The electron-beam falls on a phosphor coated, broad end (or the screen) of the tube. The tube has an electrostatic or a magnetic system for deflecting the electron beam. The broad end of the tube or the screen is coated, inside with a phosphor, which is a fluorescence producing material. Phosphor coat gets illuminated on striking by an electron-beam, and produces image, visible on the outer surface of the glass tube.

These cathode-ray tube based visual devices have, extensively been used as TV-screen and computer monitors. When the entire area of the phosphor screen is scanned in a fixed pattern, an image is produced on the screen. In a color TV, the color intensity and brightness of thus produced image is controlled by three electron beams. These beams also control the three primary colors, which are red, green and blue. An oscilloscope is a slightly different version of CRT, which produces a line diagram, rather than a picture.

3.4 Phosphor Coating and its Persistence

Phosphor coating consists of a material which produces fluorescence, when an electron beam strikes it. Various phosphors are now available to meet the needs for different measurements or display applications. The brightness, color, and persistence (duration) of the illumination depends upon the type of phosphor used on a CRT screen.

Phosphors can have their persistence timings ranging from less than a microsecond to several seconds. For visual observation of brief, transient events, a long persistence phosphor is desirable, but for events which are fast and repetitive, or high frequency, a short-persistence phosphor is generally preferable.

3.5 Color CRTs

As mentioned earlier, a color CRT uses three primary spectral colors (red, green and blue). Respective phosphors are present on a color TV-screen. Other colors can be produced by a judicious mixing of these primary colors. There are three electron guns, one each for these three primary colors. The phospors are packed in strips or in clusters, which are called triads. These are arranged, either in a straight line or in a triangular configuration. The electron guns (beams) are usually constructed as a single unit. A grille or mask is used to absorb electrons that would otherwise hit a wrong phosphor.

In case of a shadow-mask CRT, a metal plate with tiny holes, is placed in such a way, that the electron beam only illuminates the correct phosphors on the screen of the tube. There is another type of color CRT, which uses an aperture grille to achieve the same result.

3.6 Health Hazard of CRT: Ionizing Radiation and Radiation Toxicity

CRTs emit a small, but a finite amount of harmful, high energy radiations e.g. ultra-violet rays and X-rays. These can be harmful to the viewers, particular for the children, who view TV-programs for long time, from a close range. Since the children do not understand such implications, related to such radiations and health hazard, which they produce. Hence the 'Food and Drug Administration' (FDA) in the USA, has strictly limited the exposer to these radiation to 0.5 Milli-Roetgens (a unit for measuring radiations), per hour at a distance 5cm from the TV-screen. No such, radiation exposer limit has yet been fixed in India, but fortunately, all over the world, the CRTs are now getting replaced by LCDs, which are much safer.

3.7 Demise of CRT Technology, and Cause for its Demise

Although CRT has been the mainstay of display technology for over a century now, the demand for CRT's has dropped to a perceptible extent since the beginning of year 2000. This falloff has further accelerated in the latter half of the year 2010. The rapid advances in LCD technology, coupled with fall in the price of LCD's have been the reason for the demise of CRT technology.

Flat panel LCD, which was first used for computer monitor, has also been followed by its introduction into TV-sets. This has been the key factor in the demise of CRT-technology. Major companies e.g. Sony and Mitsubishi (both from Japan), have stopped CRT production since the year 2010. This has meant almost complete erosion of the CRT's capability. In the American-Continent, production and sale of high-end, CRT TV's, of 30-inch screen, ended by year 2007. This was followed by stopping the production of less expensive, combo CRT-TV's, with a 20-inch screen, with an integrated DVD player.

The demise of CRT, however, has been happening more slowly in the developing countries like India, where there has been a general reluctance, to dump off a good, working model of CRT-TV, in favor of LCD.

In spite of many improvements, CRT's have remained relatively heavy and bulky. Their space occupancy is much more space, as compared to a LCD unit. Power consumption of CRT is also high. CRT screens have much deeper cabinets compared to flat panels and rear-projection displays for any given screen size. Hence it becomes impractical to have CRTs, larger than 40 inches screen. The CRT disadvantages became especially significant in light of rapid technological advancements in LCD and plasma flat-panels which allow them to easily exceed 40 inches size limit of CRT. Large LCDs, being thin are mountable on a wall.

When we refer to the 'demise' of CRTs, we specifically mean their use as visual monitor for computer and TV. In several other applications, e.g. X-ray tubes and many physics instruments, which produce high energy radiations (including an electron beam), the use of CRT is likely to continue.

3.8 Discovery of Liquid-Crystal (LC) and Liquid-Crystal Display Technology

In the year 1888, an Austrian-Botanist, named Friedirich Reinitzer, at the Charles University, at Prague, was experimenting with benzoic acid ester of cholesterol, when he observed a distinct color effect on cooling this compound, near its freezing temperature. He also observed that the benzoate-ester of cholesterol did not melt like most other organic compounds do. It showed two different melting-transition temperatures, at 145.5°C and 178.5°C, respectively.

It produced a cloudy melt at 145.5° C, which turned into a clear, transparent liquid on further heating to 178.5° C. This phenomenon could be reversed on cooling the molten compound. Reinitzer also described the ability of cholesterol-benzoate to rotate circularly polarized light, which was a phenomenon earlier known to be associated with quartz crystal.

When this observation of Reinitzer was referred to a German-Physicist, Otto Lehman, he re-examined the phenomenon, and he found that at 145.5° C the cloudy liquid, still had crystallites. These, Lehman attributed to an intermediate fluid-state, which retained crystalline properties.

With this observation, a question now arose– "Can certain liquids exist, having a definite arrangement of molecules, just like the crystalline solids?"

Findings of Reinitzer and Lehman were reported at the Vienna Chemical Society Meeting, on third May, 1888.

This was the first reported discovery of a 'liquid-crystal', which was followed by many more of similar observations. It was much later, that is in the year 1991, that Pierre-Gilles de Gennes was awarded Physics Nobel Prize for—"Discovering that the methods developed for studying certain phenomenon's for crystals, can be generalized and applied to more complex form of matter, particularly to the liquid crystals and polymers".

Numerous examples of LC's can be found both in the nature. Technical applications of Lipotropic, LC-phases are also abundant, even in the living systems, e.g. many of the proteins, cell membranes and tobacco-mosaic virus are some of the typical examples of natural LC. Other familiar examples of LC are soap and detergent solutions, where molecules can arrange to form micelles.

3.9 What is a LC?

The earliest discovery leading to the development of LCD technology was the discovery of liquid crystal itself, and dates back to the year 1888. At first thought, the term 'liquid-crystal' appears to be a misnomer. We know that only the solids, having a regularly repeating arrangement of molecules are crystalline. LC is a state of matter that has properties, in between ordinary liquids, and those of crystalline solids. Thus a liquid-crystalline material can have fluidity, similar to a liquid, yet in the liquid state its molecules remain oriented in a regular, crystal like fashion. The structure of a typical molecular species, N-(4-Methoxybenzylidene)-4-butylaniline (MBBA), having LC properties is shown in **Fig.-2.**

A LC is not a direct light-emitting device, but uses its light modulating property. There are different types of LC-phases. These are distinguished by the difference in their optical properties e.g. bi-refringences. When viewed under a microscope using a polarized-light source, different liquid-crystal phases will appear to have distinct textures. These contrasting areas in their texture correspond to various domains, where the LC-molecules are oriented in different directions. However, within a particular domain, the molecules are well ordered. Just like water, which besides being a liquid can also exists in solid (ice) or gas (steam) phases, the LC-materials may not always be in a liquid-phase.

3.10 Classification of LC's

LC's can be classified as having 'thermotropic', 'lyotropic' and 'metallotropic' phases. Of these, the thermotropic and the lyotropic LC's are exclusively organic molecules. Thermotropic LC's exhibits a phase-transition into the LC-phase as the temperature is varied. Lyotropic LC exhibits phase transitions as a function of both temperature and concentration of the LC molecules in a solvent (e.g. water). A metallotropic LC is composed of both organic and inorganic components in a molecule and their LC-transition depends, besides the temperature and concentration, also on their inorganic is to organic composition ratio.

Besides the use of LCs in LCD, the temperature sensing and color changing abilities of liquid crystals make them suitable for use in a wide range of electronic thermometers and novelty applications such as the mood-rings of clothing's, and advertising specialties. Liquid crystal color can change through the visible region of color-spectrum with a temperature change of only two degrees centigrade. By employing variable formulations, their temperature range can be expanded and the starting point of a particular range can be changed. Conversely, liquid crystals can be made to hold a particular color over temperatures ranging from 10°C, to 60° C.

3.11 What is a LCD Monitor?

Many, modern electronic-display devices, other than TV and computer monitors, became LC-based, quite earlier. It was much later, i.e. around the year 2000 that LC-based monitors for computers and TV-screen were introduced.

It was mentioned earlier that, while the CRT based monitors were heavy, bulky and had a fragile glass-tube with a thick screen. In contrast, LCD monitor which is based on a thin film of LC-material, spread on a thin plastic panel is non-fragile and lighter.

There are certain merits of a LCD monitor, which do not escape the observation of a common user, when compared to old TV-monitors, with the visual display device based on a CRT. Thus LCD-monitors are lightweight and more compact. These are easy to shift for installing at an alternate place. Besides being compact, these are less expensive, more reliable, and more comfortable to the eyes, when compared to CRT-display. They are available in a much wider range of screen sizes compared to CRT. LCD's are now being extensively used and these are fast replacing CRT as visual display devices.

3.12 Liquid Crystal Display Device

Each pixel of an LCD consists of a layer of molecules of LC-material, which are aligned between two transparent electrodes, and two polarizing filters. The axes of transmission of these are perpendicular to each other. Since there is no LC between the polarizing filters, light passing through the first filter would be blocked by the second (crossed) polarizing filter. In most of the cases the liquid crystals have double-refraction property.

The surfaces of the electrodes, which are in contact with the liquid crystal material, are so treated, as to align the liquid crystal molecules in a particular direction. This treatment typically consists of applying a thin film of polymer, which is unidirectionally rubbed on a screen, using a cloth. The direction of the liquid crystal alignment is then defined by the direction of rubbing. Electrodes are made of a transparent conductor, made of indium-tin oxide (ITO).

3.13 Functioning of LCD

LCDs used for TV screen and computer-monitor are electronically modulated, optical devices, which are made up of a number of pixels, filled with liquid crystals and arrayed in front of a light source (black light) or a reflector, to produce images in color or monochrome (black and white).

When no electric field is applied, the orientation of LC-molecules is determined by their alignment at the surfaces of electrodes. Twisted, nematic device is the most commonly used liquid crystal device. The surface alignment directions at the two electrodes are perpendicular to each other. Under such condition, the molecules arrange themselves into a helical, or a twisted structure. Such alignment reduces the rotation of the polarization of the incident light, and the device appears slightly gray in color.

If the applied voltage is large enough, the liquid crystal molecules in the center of a layer are almost completely untwisted and the plane of polarization of the incident light is not rotated as it passes through the liquid crystal layer. This light will then be mainly perpendicularly polarized to the second filter, and thus be blocked, resulting in pixel to appear dark (black). By controlling the voltage applied across the liquid crystal layer in each pixel, light can be allowed to pass through in varying amounts thus constituting different levels of gray (dark). Thus the electric field also controls (reduces) the double refraction properties of the LCs.

The optical effect of a twisted, nematic device in the 'voltage-on-state' is far less dependent on variations in the device thickness, than that in the 'voltage-off-state'. Because of this, these devices are usually operated between crossed polarizer's, so that they appear bright with no applied voltage. The human eye is much more sensitive to variations in the dark state than that of the bright state. These devices can also be operated between parallel polarizer's, in which case the bright and dark states are reversed.

3.14 Merits of LCD

By the year 2008, worldwide sales of television sets with LCD screens, and LCD based computer monitors had far surpassed the sale of CRT units. LCD's are more energy efficient and offer safer disposal for discarded units, compared to the CRT units. Lower electricity power consumption of LCD, enables it to be used in many battery powered, electronic equipments, e.g. digital clocks), electronic calculators, and mobile-telephones. With an attachment, a LC-based visual device can have dual functions, e.g. it can function as a TV as well as a computer monitor.

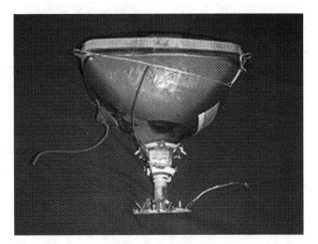

Fig.-1 Common CRT used in computer monitors and television sets.

Photo—Blue tooth, Taken from Article on CRT
http://en.wikipedia.org/

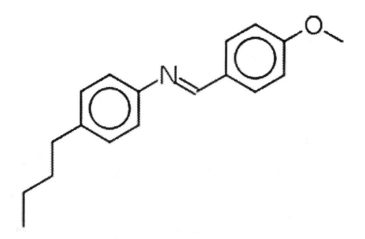

Fig.-2 Chemical Structure of N-(4-Methoxybenzylidene)-4-butylaniline (MBBA) Molecule, which forms Liquid Crystal.

Photo—J Kwchui, Taken from Article on CRT http://en.wikipedia.org/

Chapter-4

❖

Conventional Photography *verses* Digital Photography

4.1 Introduction: Pre-Dawn, Natural Photography

Activities, which were somewhat similar to 'photography by a camera', has been inadvertently taking place in nature. One such incidence, which has been well recorded by the famous painter, Leonardo da-Vinci (who painted the well known portrait of Monolisa) is being mentioned here. Leonardo has mentioned about the presence of a small hole in the wall of a dark cave somewhere in Europe, which acted like a pinhole-camera. This small hole, laterally projected an inverted image of out-side, sunlit valley on a piece of paper, which has been lying in the cave. The paper got yellowed due to long exposer to light, producing an inverted image of the valley. This was an example of a natural photography, which happened much before the technology of today's photography was developed. Cameras with lens, to form well defined images were designed much later.

It appears that the inventors of modern photography were basically interested in developing some means, to quicken the process of imaging an object. The overall process of the photography is one, in which an image formed by a camera, which we know as a photo-print should be 'fixed' and retained permanently.

Another early example of producing fixed images (photographs) has been due to a person, Josiah Wedgwood (he was a potter by profession). Wedgwood made an attempt to reproduce certain images even without using a camera. He obtained copies of certain paintings on leather sheets, which was based on blackening of silver salts, on exposer to light. However he could not fix the prints, because he had no means to stabilize the image. Without washing away of the unexposed silver salts, the images which he produced ultimately turned completely dark (black) in light.

4.2 Invention of Modern Photography

The term 'photography' is derived from two Greek-words, i.e. φῶς, which means photo, or an image and γραφή or graphe, which means recording with the help of light. Literary it means, 'representing by means of light' or 'drawing using light'. The process of inventing and improving photography, using camera, started much later i.e. in the first decade of nineteenth century. At that time it was a prevalent practice of painting portraits by skilled painters by hand. Only the very rich persons could afford this; and the development a new photographic technology was desired to substitute the then practice of painting of human portraits.

Development of the technology of photography, as we know it today, has been a long and slow process; and a large number of persons made many significant contributions towards the development of photographic process.

In the year 1816, Johanna Heinrich discovered that a mixture of very finely divided metallic silver and chalk powder, darkened on exposer to visible part of light spectrum. Later, in the year 1822, photo-etching to produce an image was carried by a French inventor, Nicephore Niepce. Soon after this, in the year 1825, Niepce made his first permanent photograph of natural scenery, which required eight hours exposer. To shorten the exposer time, Niepce sought the collaboration of a fellow scientist, Louis Daguerre. Together, these two scientists/ technologists started the use of silver halides, (AgX, where X=Cl, Br or I) as the photosensitive materials. It was in the year 1838, that Daguerre successfully took his first photograph of a pedestrian in a Paris street, while having a shoeshine, at that particular time. It took several minutes exposer to take the photograph. Daguerre's invention of photography was announced, as a major scientific achievement by his country and as a gift to the whole world from France.

Somewhat earlier to this, i.e., in the year 1832, Hercules Florence devised a very similar process, which he named 'Photographie'. Yet another English inventor, named William Fox Talbot discovered a somewhat different process of fixing an image on paper. His method was also based on the use of silver halides, but the details of the process were kept secret.

On knowing about Daguerre's work on photography, Talbot made further improvements in his process. Due to all these improvements, making of personal portrait, became a process, readily available to the masses, and at a very reasonable price. Talbot had invented a process, which produced a 'negative image' of an object. This meant that the bright parts of any object, which was photographed, appeared darker and *vice-versa*. Most famous of Talbot's photography has been the print of the 'Oriel-Window' in Leacock Abbey. This has been known to be the oldest (1835) photo-negative **(Fig.-1)** in existence even today.

4.3 Basic Principles of Photography

The technology of conventional photography, known as 'black & white photography' is based on, initially creating of a 'latent image' of an object, to be photographed, on a photographic film or paper. The term, latent image means that the image is not as yet visible to eyes, but it is a 'real image' (according to physics), i.e. an inverted image of the object, which has been photographed. A latent image is formed due to the action of light (radiation) originating from the object, on a thin layer of a photosensitive material (silver halides) on a film. For photographing an object, the light reflected from the object has to be properly focused through a lens, so that it can form a very sharp latent image of that object, on a light-sensitive surface.

Recording of such an image (photograph) is carried on a light-sensitive surface, which can be that of a transparent (celluloid) photographic film-strip / glass-plate, or a photosensitive paper, which has been coated with a photo-sensitive material, e.g., a judicious mixture of silver halides.

In conventional photographic process, radiations, which form a part of the visible spectrum, act upon the light sensitive substances e.g. silver halides. Thus sensitized silver halides, become more prone to getting reduced to metallic silver by a weak reducing agents (a developing agent, hydroquinone, metol and bisulphite and a mixture there of), which is called a 'developer'.

4.4 Photographic Camera

A 'photographic camera', is sometimes called a 'camera obscura'. The term 'obscura' means devoid of light. It is a dark chamber from which, the light is completely excluded. In the dark chamber of a photographic camera, there is a provision for controlled admitting of light, *via* an aperture (opening) and a lens. A converging (convex) lens helps in proper focusing of an image of the object to be photographed. For controlling an exposer, i.e. the time of light falling on the film, a shutter, with a variable exposer time is provided. The object to be photographed must be variably illuminated, depending upon the desired color contrast of a photograph to be produced.

Size of cameras can range from very small to very large, in size. Even it has been possible to have a room-sized camera, which is kept dark, while the object to be photographed is kept in an adjoining room, where it is properly illuminated. This type of arrangement was commonly used for 'reproduction photography' of a flat copy, when large film-negatives were used.

However it is a general principle, known since the birth of photography that—'Smaller is the camera, brighter and sharper

46

shall be the image'. It may also be mentioned that soon after introduction of, the technology of photography, extremely small, detective cameras were made available. These could be fitted into a wrist-watch, a coat-button, or a tie-pin. These were the devices, which were somewhat similar to those we know today as 'close-circuit, TV-cameras'.

4.5 Modern Cameras

An advanced camera, used in today's photography is a device, which can cause a time variable exposer and a good focusing of an image of an object, with the help of a lens system. A changeable, telescopic lens can be used for focusing distant objects. The exposer time can be varied from fraction of a second to several minutes. The result of such an exposer produces a latent image of an object, on the light-sensitive film or paper placed inside the camera. No light can enter the camera, unless its shutter is opened to produce a pre-determined exposer to the well focused object, which is meant to be photographed.

In some old models of cameras, the focusing mechanism was based the variable length of a bellows, which could be adjusted. In modern cameras a lens, which looks like the concentric sliding tubes in a telescope, can be moved to adjust focus. Modern cameras **(Figs.-2)** are also provided with photocell based, 'exposer-time meter', which can measure light intensity and accordingly suggest to an amateur photographer, suitable exposer time. For photographing distant objects, the cameras can be fitted with telescopic lens. Near sea-shore and snow-clad mountains, where there is a high intensity of UV-radiations, UV-filters for camera lenses are used to diminish UV-intensity, which can cause over-exposer of the film.

4.6 Light Sensitivity

Variable light sensitivity of photographic films and papers arises due to a coating of a judicious mixture of finely divided silver halides (chloride, bromide and iodide) on them. These silver halides are prepared in complete darkness and are coated on transparent celluloid or plastic films, glass-sheets or papers, with the help of a suitable adhesive, e.g. gelatin. All these manufacturing operations, used in making films and printing papers are completely automatic, and light is completely excluded during the process of making these. After manufacturing, the photographic film and printing-papers, thus produced, are wrapped and stored in light-proof packaging. An expiry time period is mentioned on the packages, before which the photographic material must be used.

1.7 Photography as a Profession

Professional photography has several uses for business, art, and recreational purposes. It has applications in medicine (X-ray photography), science and manufacturing processes e.g. photolithography and printing. Final image of any photographed object on a paper-base, known as a 'print', and this is generally what a customer ultimately gets from a professional photographer. Using light sensitive dyes of different shades and color, it has been possible to make 'color-print' by the conventional photographic process.

This is not the end. By continuously photographing a moving object, on a reel of film, after small time intervals (few seconds), it has been possible to do a video or 'film-photography'. When a reel of a video-film, is screened continuously, it shows the object moving. This is what happens naturally, when we see an object in motion. The impression of an image, produced on human retina lasts for few seconds. If new images continue to be formed, on the retina, it gives the perception of a motion.

A professional photographer, used to have a studio and a dark-room for processing 'black & white photo-prints'. He used to be a skilled person, who could judge the lighting conditions and adjust exposer time accordingly. Frequently, there used to be a need to 're-touch' the negatives, before final printing on paper.

There used to be photographic clubs in schools and colleges, where students learnt, elementary skills of photography, including developing and printing. With the advent of 'digital photography', there is far less to learn about the 'skills', which a photographer had to learn earlier. To some extant his has taken off the business of small time, town photographers. Yet due to money-power, newer trends have, now emerged in the profession of photography, e.g. movie recording of occasions like marriages and birth-day parties. Some 'stills' can later be chosen from the movies and made into enlarged color prints for an album.

4.8 Standardization and other Photographic Processes

An inventor, Sir John Herschel made many new contributions to the process of photography. He invented a cheaper, and non-silver halide based process of copying certain prints, which is now known as making of 'blueprint'. This process, which is based on photosensitivity of a ferrocyanide complex of iron, is useful in making maps of architectural plans and technical designs for construction workers and engineers. It was also Herschel, who introduced the terms 'photography', 'negative' and 'positive' prints. In the year 1839, he also discovered that a solution of a chemical, called sodium thiosulphate, can dissolve out unexposed silver halides from a print. After this discovery, Talbot and Daguerre developed a method to 'fix', or make a print of a picture 'permanent', by removing un-sensitized and un-reduced silver halides from the printing paper. He was also the first to use silver halide coated glass plates to make negatives.

On 14, March, 1839 Sir John Herschel, in a lecture before the Royal Society of London, described the details of chemistry, under-lying photographic process, and coined the word 'photography'. Somewhat earlier to this (25, February, 1839), an article was published, in a German newspaper (Vossische Zeitung), by Johann von Maedler from Berlin, who was an astronomer by profession. He was the first to use the term photography.

By early 1839, all the process-details of 'black & white' photography, based on the light sensitivity of silver halides, have been fully worked out. Even the chemistry under-lying the process, was being taught to science students. It was the time now, for industries, e.g. the Eastman Kodak in USA and Agfa in Germany to enter into commercial production of cameras, films and other photographic materials.

4.9 Chemistry of 'Black & White' Photography

A silver halide is sensitized on exposer to light and there is a distinct change of its color. Thus freshly precipitated, white silver chloride changes its color to light, violet-pink, when exposed to light. Light sensitized silver halides become more easily reducible by weak organic reducing agents e.g. hydroquinone and metol (another, soluble phenolic compound). Thus reduced and very finely divided metallic silver particle, which adhere to the film or paper, are no more lustrous white, but appear black. This is because these particles are too small to reflect light falling on them.

Photo-sensitivity to light radiations for different silver halides varies, and also differs in different parts of the visible spectrum (red to violet). Hence a 'back & white' photograph does not show the same depth of shades, which we see, in an object from our eyes. Photographic films and printing papers of different light sensitivity are made using a mixture of silver halides. That is why only a skilled photographer can pre-judge the results of his photos and frequently he has to do 'retouching' of a negative before making final prints.

When a properly exposed photo-film is immersed in a solution of 'developer' a negative print of the film is formed. In a negative print, the more-exposed (to light) parts become black and unexposed or slightly exposed parts remain white or slightly dark, depending on the amount of light falling on those portion, during exposer.

At this point the film cannot be taken out into light, because whole of it shall get darkened by light. Hence it becomes necessary to 'fix' it, by washing with sodium thiosulphate solution to remove unreduced silver halides. Professionals know this chemical as 'hypo'. All these operations, which are carried in a dark room, produce a so called 'negative', where the darker portions of the photographed object appear bright and *vice-versa*. The whole process is now repeated, making the negative (film) as the object, to take its print on a photographic paper.

Durability of silver halide based 'black & white' photographic prints is due to the inertness of metallic silver, which is a 'Nobel Metal'. It is not unusual to find in good condition, very old family photos in many houses, which were taken more than a century back. In contrast silver halides based black& white prints the, color-photographs have a tendency to fade away or get disfigured due to long exposer to light and air.

When color photography was introduced, it was costlier, but ultimately it became cheaper, due to the lower cost of material (photo-sensitive organic dyes), it used. 'Black & White' photography consumed a sizable portion of silver produced, world-wide.

4.10 Video and Audio Recording on Magnetic tapes

During the year 1950 onward, video, audio and computer data recording, using analog-system on plastic tapes, coated with magnetic material emerged as a powerful technology. However this proved to be a short-lived technology in at least some fields and where ever possible, it has now been replaced by the digital technology. More details about recording on magnetic tapes have been described, elsewhere **(Chapter-7)** as a separate chapter in this book.

4.11 Emergence of Digital Photography

Until the advent of digital technology, the conventional photography was always done by exposing light sensitive photographic film or paper and used chemical processing to develop and stabilize the image. In contrast to this, digital photographs can be displayed directly with the help of a computer and printed on ordinary paper. It can also be stored for reuse on a CD, and manipulated for transmission to another computer. Digital photographs can be archived using digital-computer techniques, and does not need chemical processing at any stage. In contrast to the 'conventional photography', which is based on the action of light on, photosensitive film or paper, the 'digital photography', is a form of photography, which uses an array of 'light sensitive sensors'. There are two main types of sensors, which are-

1. Charge-coupled device (CCD). In these a photo-charge is shifted to a central, charge-to-voltage converter.

2. Active Pixel Sensor or CMOS sensors. These sensors capture an image, focused by the lens of a 'digital camera'.

The image, thus captured is stored, just like any other digital file. This file can be processed and printed, with simultaneous color correction and enlarging or reducing in size. The picture file, thus produced, is readily available for viewing or digital printing.

4.12 Digital Camera.

Developments in the technology of digital cameras have been very fast. Some digital cameras have a built-in, solid state flash memory, while others have a removable one. Digital, tapeless camcorders (as these are called now) can function as 'digital still camera' using flash memory-discs, and internal hard drives. Certain old models of digital cameras, e.g. 'Sony Mavica' range model used removable floppy disks and mini-CDs. Simple and cheaper web-cams have a digital memory device (memory card) for storing images, which can also be transferred to a computer. More sophisticated cameras have linear array type digital memory.

Besides taking still pictures, Digital cameras **(Fig.-3)** can also simultaneously record sound and videos. Some cameras can be used as webcams. Others can be used as 'PictBridge' standard, to be connected to a printer without using a computer, while some others can display pictures directly on a television screen. Many camcorders, after taking still photographs, can store them on videotape or on flash memory-cards with the same functionality as digital cameras.

Just as in case of film cameras, the quality of a digital image depends on many factors. Most important of these is 'Pixel-count', which is generally listed in 'megapixels'. From the Pixel-count, a camera user can evaluate the capabilities of his camera. Besides Pixel-count, it is the processing system inside the camera, which turns the raw data into a color-balanced and pleasing photograph.

4.13 Conclusion and Final Word

As a final word, digital cameras and photography have completely changed the process of conventional photography. Photography has now become much cheaper and does not need more training or expertise. This has almost phased out the conventional photographic process and the conventional old cameras are now becoming antique pieces.

However some experts have opined that the quality of digital video and audio, to some extent lack the quality attainable by the conventional methods.

Fig.-1. A latticed window in Lacock Abbey, England, photographed by William Fox Talbot in 1835. This is the oldest known photographic negative made by a camera.

Photo-William Fox Talbot, taken from Article on Photography
http://en.wikipedia.org/

Fig.-3 Nikon F (959), a 35mm film camera
Photo—Jeff Dean, taken from Article on Photography
http://en.wikipedia.org/

Fig.-6 A modern digital camera.
Photo—Jeff Dean, taken from Article on Photography
http://en.wikipedia.org/

Chapter-5

—◈—

Typewriter *verses* Personal Computer

5.1 History and Invention of Typewriter

In the year 1714, Henry Mill patented a writing machine, which can be regarded as the precursor to a modern typewriter. Just like in normal writing by hand, using ink, pen and paper, Henry's machine was capable of transcribing on a paper, any desired material to be written. Thus, using Henry's the whole matter, which was desired to be written, got transcribed on a sheet of paper. Transcription of the write-up by Henry's machine was so well and neat, that it looked, just like any material printed from a press. Patenting of his machine, by Henry was followed by a number of other inventors, who made somewhat similar writing machines, with increasing and improvements and sophistications.

Thomas Alva Edison (see also **Chapter**-2) has been credited to the invention of several machines. In the year 1870 he got still one more such credit, when he designed a workable, key-board typewriter. His typewriter could be operated manually as well as by electricity.

5.2 Need for a Writing Machine

The invention of a writing machine or the typewriter coincided with the time, when the business communications were fast increasing all over the worlds. Hence, there was a necessity being felt to develop a mechanized writing machine. Such a machine could be operated at personal level, by any person. This was also the time, when a well trained stenographer could take a dictation, in a well developed, 'short-hand script' at a rate of about, 130 words per minute. Yet, during that vary period, even a good writer, using a conventional pen, ink and paper could at the most write only about 30 words per minute. This was being considered as a big handicap for office work in government and private offices and for the freelance literary writers of that period.

Presses have already been established at that time for printing hundreds of copies of a document, or for printing of books. This happened earlier to the designing of s writing machine, i.e. a typewriter. But, when a person wanted to have only a few copies of a document to be prepared, for his personal or office use, then the only way left out was to make, required number of hand-written copies, either personally or with the help of another person (copier) with good handwritings.

After the invention of such writing machines, the name 'typewriter' was coined for them. Since the time of its invention, and start of its industrial production in the late nineteenth century, typewriters have become indispensable office equipment, as well as for personal use.

5.3 Improvements in Typewriters

During the period 1829 to 1870, many typewriter machines were developed and patented by several inventors and companies in Europe and in USA. However none of these machines were produced commercially. It was in the year 1873, that a prototype of a typing-machine, was made by the 'Sholes and Glidden Typewriter Company', which was the first, commercially successful typewriter, to be manufactured as well as marketed.

From the very beginning of commercial machines, the sequence and placing of various letters on the key-board of a typewriter, was determined from the general frequency of their occurrence in any commonly written text. Naturally it was different from the sequence of their common placement as Roman-Alphabets, i.e. A, B, C, D, F, etc.

The convenience in using various fingers, of both the hands, of an operator, placement of different keys, on the key board was also taken into considered. Sholes and Glidden Typewriter was the first machine to have used, 'Qq, Ww, Ee, Rr, Tt, Yy— etc., as the sequence of letters (capital and small), starting from the top second row of the letters on its keyboard. This sequence of letters, has not only remained unchanged even today, but has also been adopted in the modern, computer keyboard.

Remington Company, which had earlier been known as the manufacturer of sewing machines only, soon commercialized a machine, which they called, 'Sholes and Glidden Typewriter'. This has been the origin of the term, 'typewriter' as a word for the machine, as we know it today (Fig.1). Remington manufactured its first typewriter in March, 1873, in its factory in New York City. It had the same, ('Qq, Ww, Ee, Rr, Tt, Yy—'), layout of letters on its keyboard.

Initially the typewriters were manufactured as, large 'table-model' only, but looking to the need of travelling persons, portable models were also introduced. In case of electric typewriter, a provision was also made for visual display of one or two typed line, and correction therein, before it is finally printed on a paper.

5.4 Designs and Developments of Typewriter

By the year 1910, the 'manual' or the 'mechanical' typewriters, manufactured by different companies had reached to a somewhat standardized design. A typewriter can be mechanically (manually) operated or it can also be an electrically operated machine. When the key of a typewriter for a particular character, i.e. alphabet, digit and punctuation mark or any other marks etc. is pressed, it causes that character to be printed on a paper inserted into the bar of the machine. Each character, via an inked ribbon, makes an impression on the paper. These typed characters are similar to the cast-metal type-pieces, being used in a printing press.

More papers, each one with a carbon-paper, can also be inserted into the bar of a typewriter, if copies are desired. When a large number of copies are desired, these can be made using a typewriter, and a specialty made 'typewriter-stencil', made for this purpose. By striking a key little harder, the desired script is typed on the stencil-paper, by cutting off out-line of the characters on the stencil-sheet. Desired numbers of copies are then made using the cut-stencil and a duplicating machine.

5.5 Operation of a Typewriter

Most manufacturers of typewriters, followed a general concept that each key was attached to a 'type-bar', which had a particular character (capital and small alphabet) molded, in a reverse (mirror-image) form, into its striking head. When a key was briskly struck, the type-bar hit on an inked, 'ribbon' making a printed mark on the paper, wrapped around a 'cylindrical platen' (bar). The ribbons, which was about a centimeter wide, was made from fabric, and these were inked in 'black only', or sometimes in 'black and red' stripes, each being half the width of the entire length of the ribbon. A lever on the typewriter machine allowed switching between these two colors. Use of two colors on the ribbon was considered useful for high-lighting certain entries. The platen was mounted on a carriage that automatically moved to the left or to the right, for advancing the typing position horizontally, after each character was typed. The paper, which was rolled around the typewriter's platen, was then advanced vertically by the 'carriage return' lever, at the far left, or sometimes on the far right, into a position for each new line of text.

5.6 Improvements in Typewriters

For any invention, the technicians working on its manufacturing line have made important contributions, for its improvement and innovation. This has happened in case of most of all the invented machines, e.g., automobiles, television sets, gramophones, microwave ovens, telephones by no exception. Such innovations have resulted in ever-more, successful designing of new commercial machines.

This has also been true, in case of typewriters. Typewriters have been made, throughout the world, but most of these were based on more or less similar designs. Numerous designers, belonging to different companies of the world, have worked for decades to make incremental improvement of their machines. Thus the earlier typewriters were invariably large and table models **(Fig.1)**, but manufacturers soon started making 'portable typewriters', which proved to be useful for traveling press-reporters.

Typing errors have been a major handicap, because just like a hand-written document, there was no way except to strike off the mistake, which still showed on the paper. In case of some advanced, electrically operated typewriter, one or two typed lines at a time, are visually displayed, before these are finally typed on a paper. This provided a somewhat limited facility for a correction, before the matter was finally put on the paper. However it created, some extra burden for the typist, because each time he has to look for the correctness of the visually displayed line, before transferring it on paper. This also resulted in slowing down the typing speed.

5.7 Need for Trained Stenographers and Typists

Generally there have been separate posts of 'stenographers cum typists' in large government or private offices. A stenographer is trained to take a dictation in 'short-hand', which he can, then type on a typewriter. 'Hindi short-hand' has also been developed, looking to the demand of such manpower, in India. Typewriter with Hindi alphabets were also manufactured. As mentioned earlier this was possible because Hindi, written in 'Devnagari-Script' is adoptable in typewriting. Unlike typing in Hindi, using Devnagari-script, a typewriter could not be made to adapt to the 'Persian-Script', which is being used in writing in 'Urdu' language, being used in India and Pakistan.

By the end of 1980s, and with the introduction of 'word-processors' and 'personal computers', office typewriters are now fast becoming items of past. These have become items of curiosity for new generation, and these can still be seen lying isolated in remote corners in offices.

5.8 Manufacturers of Typewriters and their Decline

Remington, Underwood and Smith-Corona Companies have been the most important typewriter manufacturers in the world. These companies captured most of the Indian market as well as the markets of many of the developing countries for their requirement for typewriters. Godrej & Boyce was the first company in India, which after independence started manufacturing typewriters. Soon it became a familiar name among typewriter manufacturers. Just like many other multinational companies, Godrej has now closed down its typewriter-manufacturing unit.

5.9 Ode to the Typewriter

Earlier to the introduction of personal computers, typewriters have become a very favorite tool for writers and journalists. As a farewell to the typewriter, some journalists have written, sort of 'condolence articles' in newspapers and magazines, when the typewriters were phased out. Thus an article, "Ode to the Typewriter" by Bachi Karkaria (a Mumbai based journalist) was published in the 'Times of India' (May 15, 2011), while another article to bid 'Farewell' to typewriter appeared in, 'India Today, Weekly' of May 16, 2011. These articles exhibit a very personal attachment of these users to a typewriter machine.

5.10 Personal Computer Replaces Typewriter

Personal computers were not specifically designed to replace typewriters. A personal computer (PC) is a general-purpose computing facility **(Fig.-2)**, which can be availed by an individual or a family at their home. It is also useful equipment for office use. A personal computer does not need a highly trained operator to help any user. A person, who does not have a formal training of computer, can also use it as a writing machine and many other applications. A computer keeps on giving necessary instructions to the user.

The size of a PC and its cost-price becomes an important consideration for the end-user, who wants to own it. Fortunately PCs have now become affordable to a common person. There is a general trend among persons (including those in India), who own a PC to frequently go for a more advanced model of PC, and this has created a huge market for 'second-hand PCs', which are still in good working condition and can be resold. In India a PC has now become a routine home facility (while typewriters were not), but still there are scattered, rental as well as 'on the spot', typewriter and PC services available in any medium sized city and town in India.

5.11 Advanced Computers

In contrast to a PC, there are more advanced computers, used in batch-processing and time-sharing systems, which can be used by several persons, at the same time. These machines are generally installed in business houses, industrial establishments and science / technical centers. These require full-time operators and can do large amount of data-processing. Since we are only concerned here with the use of a PC as a substitute of typewriter, other applications of advanced computer or a PC are not being discussed here.

5.12 Laptop and Notebook

Personal computers are now available as 'Desk-Top' models, as well as 'Carry-With' type models. The later are generally referred as 'laptop', or 'hand-held' models. Smaller models of these are called 'notebook'. The laptop is a self contained one piece model; where as the desk-top models of PC's are generally three piece equipments. These three components are referred as-

1. The 'Central Processing Unit' or the CPU',
2. A 'Monitor', which is a 'Visual Display System' for any written draft, computational material (language) as well as pictures and diagrams, and
3. A 'Key-board' or the 'Programmer' also called the 'Drafting System', which is similar to the key-board of a conventional typewriter, but having some additional, operational keys.

A laptop combines all these three components in one piece. In place of keyboard, a laptop has 'touch-screen' system of operation.

5.13 Limitations of a Typewriter: Merits of PC

Personal computer can do a lot of computational work, but in this article we havel only confine ourselves to the use of a PC, as a 'better' substitute of a typewriter. The alphabets, which a typewriter can engross on a paper, are confined to 'capital' and 'small' letters only, and in one size. In case of a PC, one can have the choice of writing in alphabets of different 'size' and different 'font'. These fonts are designated as 'Times New Roman', 'Arial' etc., as well as these can be in 'italics' and 'bold'. Additionally there is a choice of 'line-spacing', 'alignment' and 'indentation'. Earlier, these were the normal facilities available only in 'printing-press', and most of these facilities were not present in a typewriter.

In case of a typewriter, a misprint cannot be corrected easily or erased, and the manuscript cannot be stored for reuse or later use, while this is possible with a PC. PCs are provided with such 'tools' as 'spelling and language correction', hence the user does not have to refer frequently to a dictionary. After completing the writing of a manuscript on a PC, one only needs to have a compatible printer to make desired number of copies.

5.14 Microprocessor and Software

Most of the PCs use the common software for processing which are 'x86-compatible' CPUs. The software applications for personal computers include word processing, spreadsheets, databases, web-browsers and e-mail clients, digital media playback, computer-games, and myriad personal productivity and special-purpose software applications. Modern personal computers often have connections to the 'Internet', allowing access to the 'World Wide Web' and a wide range of other resources.

5.15 Conclusion

A PC has a good storage capacity and modification facility for the documents prepared on it. This is quite unlike a draft prepared on a typewriter, which cannot be stored, recalled, changed and redrafted. Basically the PC was never, developed as a simple replacement or to substitute a typewriter. Though one of its important functions has been as an improved, substitute, for the 'retiring office tool' i.e. the typewriter, but it has became a multipurpose house-hold tool.

Fig.1 A Modern Typewriter
Photo, SK Milan, taken from Article on Typewriter,
http://en.wikipedia.org/

Fig. 2. A Table Top or Personal Computer
Photo, B. Boffy, taken from Article on Typewriter,
http://en.wikipedia.org/

Chapter-6

Gramophone, Records, Magnetic-Tapes and Compact Discs for Audio-Recording and Playing

6.1 Introduction

Even a primitive man might have desired to hear the voices of his ancestors. He may also have desired, to hear over and over again his own voice, songs and music as well as those from his near and dear ones. Such a desire might have made the modern man to work on invention and development of technology, to design a system for audio recording, and replaying of such recorded sound.

Sound arises due to the mechanical vibrations, produced in air, which is the media for the propagation of sound waves. Vibrations due to sound waves are very soon dissipated and a particular sound is no more heard. In recording of any sound, a lot of consideration has to be given to the pitch, which is determined by the frequency of sound waves and the loudness or the volume of the sound to be recorded. With the development of science and technology, efforts to understand sound vibrations, started right in mid-nineteenth century. This followed the development of sound recording and audio playing technologies.

6.2 Historical

In the year 1857, Édouard-Léon Scott de Martinville, a citizen of France, designed a device, which was based on a pen which vibrated due to sound waves. This was used to graphically represent human voice on a sheet of paper. Thus produced representation voice was typical for each person. Such a sound recording device was called 'Phonoautograph', which means 'signing with sound'. However, Martinville failed to develop any means of playing the voice back, and to reproduce the recorded. Besides Martinville, certain other inventers, namely Marey, Rosapelly and Barlow also created, somewhat similar devices for recording of voices, but they also could not develop any method to replay the recorded sound. Hence, the famous inventor Thomas Alva Edison has been given sole credit of inventing a gramophone or phonograph, which was the first sound recording and replaying machine. The original Edison's machine could do recording as well as replaying of recorded sound, but this is not the case with commercial gramophones, which can only replay a sound, recorded by industries. This patent of Edison's machine was to end by year 1877.

6.3 Edison's Phonograph

Thomas Alva Edison (See also **Chapters-2)** was one of the greatest inventors of his time, who designed several machines and gadgets. Much later, in the year 1899, Edison developed a unique sound recording machine, which he called 'Phonograph'. Unlike the Phonoautograph, Edison's machine was capable of both recording and reproducing the recorded sound.

There has been some similarity in names of the Martinville's sound recording device (Phonoautograph), and the later developed Edison's phonograph, which could replay the recorded sound. Besides this similarity in the names, there is no evidence which shows that Edison's Phonograph was based on Martinville's, Phonautograph. Hence, Edison has been credited to be the sole inventor of phonograph, as we know it today. There has been a major difference between Edison's originally designed Phonograph **(Fig.-1)**, in which the recording of sound and replaying it back, was done on a cylinder, from its modified version of modern phonograph **(Fig.-2),** in which flat, circular disc-records are used. Disc-records are now made by several different manufacturing companies, in USA, Europe and in India, using different materials.

6.4 Edison's Cylinder Phonograph and Disc Recording

During the early years of 1880's Edison's cylinder-phonograph, dominated the market for recorded songs. Later on, Emile Berlner invented lateral-cut, disc-records in the year 1894. Berlner started marketing disc-records, under his own label of 'Berlner Gramophone Record'. Initialy, the Berlner's records had a poor sound quality, which was later on improved by the development work carried out by Eldridge R. Johnson. He improved the sound reproduction to an extent, which was comparable to Edison's recording cylinder. Later on, these two companies, belonging to, Johnson, and Berliner respectively, were merged together to form a new company, which was named as 'Victor Talking Machine Company'. The products of this new company, dominated the market for several years.

To compete with the merits of disc-recording, Edison, in the year 1909, introduced the 'Amberol Cylinder' which had a maximum playing time of four and a half minutes, at 160 rpm. This in turn was superseded by the, 'Blue Ampere Record' whose playing surface was made of celluloid polymer. Celluloid was an early, semi-synthetic plastic, made from natural cellulose. Till then, purely synthetic, vinyl polymers were not discovered. Overall, the plastics proved to be far less fragile compared to wax, which was used earlier for making records.

The earlier patents for the manufacturing of lateral-cut, disc-records expired, by the end of the year 1918, which opened an opportunity for several other companies to manufacture sound recording-discs. This ultimately resulted in the decline of the popularity of Edison's cylinder recording machine, the production of which was ultimately stopped in the year 1929.

6.5 Emergence of Audio Recording on Magnetic Tapes and Compact Disc

Since mid-1950's, magnetic tapes became an alternative and more popular method of sound recording. Magnetic-tape recording has been particularly useful for broadcasting industries, where frequent and quick recording was required. The recording of sound on magnetic tape has been very much simpler, compared to the recording on flat gramophone discs. Magnetic tapes also had an advantage of being reusable after erasing off previously recorded sound. Along with development of tape-recording and playing back machines, gramophone and disc-records dominated the sound recording and playing market, till around early 1980's. Since then, recording by digital technology, on 'compact disc' or CD, has become yet another method of sound recording and playing.

6.6 Gramophone and Disc-Records

The word 'Gramophone' has been used, in England, for the machine used for playing, recorded sound. It is also called a 'Phonograph' in America, while a suffix 'record' with these words is used for the circular flat-discs (gramophone-record), on which sound is recorded. A disc-record is an 'analog sound storage medium' consisting of a thin, flat disc of vinyl polymer, on which are inscribed, modulated spiral grooves. These grooves, which start near the outer periphery of the disc, end near the center. This is just opposite of the grooved, CD's spiral pit (to be described in a later chapter), which starts near the center and works outwards. There is a central hole for mounting the recording disc on a circular turn-table of a gramophone.

Recording was originally done on wax coated metal surfaces. Victor Talking Machine Company started the use of circular glass sheets (discs), having wax coated with soot (carbon powder, to make the wax surface conducting) for recording. A metal replica of the record was then made by electro-deposition of copper on carbon coated wax disc. Ultimately with the help of this metallic replica, any number of plastic records could be molded.

Gramophone disc-records have been made from several different materials. The material should be hard enough, to with-stand any erosion by gramophone sound-box needle, yet be smooth and moldable. Initially a composite, based on shellac, (natural, tropical plant exudates) was used for making disc records, but later on these were made from synthetic vinyl polymers. Recording was done on both sides of a disc. Phonograph-records are generally made in 7, 10 and 12 inch diameter sizes. These could have rotational-playing speeds of 33⅓, 45 and 75 rpm. The modern disc-records can provide mono—as well as stereo-phonic sound reproduction.

Initially the audio-recording was intended solely to depict the visual characteristics of sound. It has now been realized that such depiction of sound can be digitally analyzed and reconstructed into an audible recording. As an example an early phonautogram, which was made in 1860 and which has been the earliest known audio recording existing even now, has now been successfully reproduced into sound using computer technology.

6.7 Playing of a Gramophone

During playing of a gramophone, the record is placed on a revolving turn-table. Retractable gramophone sound-box, with a smooth, changeable steel needle moves along the modulated grooved record. The vibrations, thus produced in the needle are conveyed to a vibrating diaphragm in the sound-box of the machine, which in turn produce audible sound waves in air. These are directionally magnified by a loud-speaker.

Gramophones were built in the form of a portable box, in which there used to be a built-in loud-speaker. The table-models of gramophone had an external, detachable, polished-brass loud-speaker. A picture of such a machine, having a dog, sitting in front of it, and listening to 'His Master's Voice' was used as a patented logo, by the His Master's Voice Company, which manufactured gramophone-machines and records.

His Master's Voice (HMV) has been a pioneering company, manufacturing and marketing manually operated gramophones, and records for a long time. Because of a well developed cinema industry in India, and a large market for recorded music, including those in different regional languages, audio-record making industry was establish. The industry flourished well during the pre-independence days in India. Music records of trade names HMV, Columbia, Twin-Records and Marwari-Records, were made in India, during pre-independence days. Record-playing gramophone machines were, however, imported into India. Recording of cinema music also boosted gramophone industry.

These gramophone machines were generally manually powered by a winding spring, because the supply of electricity in India, at that time was limited to big cities only. Such, manually operated gramophones could also be taken and played out-door and at any picnic spot. Production of gramophone and records has generally been phased out now in India and these are being preserved and looked upon as antique items in some families.

6.8 Magnetic Tapes, Tape Recorders and Player

Gramophone disc-records, which replaced Edison's 'phonograph cylinder', and with which it had co-existed till 1920's have been the primary medium used for popular music recording and playing, for most part of the 20th century. Its rival for audio, (as well as video) recording, i.e. the magnetic-tape recording system (See also **Chapter-7**) became popular in mid-1950s. Magnetic-tape recording has been much easier and cheaper and it could be done by an individual at home, while the gramophone disc-records were amenable only to a manufacturing company. Though, commercial, audio-recorded music is generally patented, patent infringement of commercial audio-recording become common, due to the ease with which magnetic-tape could do audio copying.

As mentioned above, sound recording and playing, using magnetic tapes emerged as an alternative to gramophone in early 1950's. Sound and many other electronic signal recording can be done on magnetic-tapes. These are plastic tapes, coated with magnetic material (oxides of iron), capable of recording sound in the form of electromagnetic signal, which can be played back into sound. These tapes, which can be one or two tracks, are available in the form of reel or cassette. These tapes record a fluctuating electrical signal by moving across a tape-head that polarizes the magnetic domains in the tape, in proportion to the audio-signal.

Unlike disc records, which have a playing time of 3-5 minutes only, a tape-cassette can record and play-back for 30-120 minutes and has proved to be much cheaper compared to gramophone audio-records.

6.9 Working of a Tape-Recorder Machine

The machine used for recording and playing-back sound, using magnetic tape, is called a 'tape-recorder'. These are electrically driven machines, based on a three motor system. One of these is a constant rpm motor, which is used to drive the capstan. Use of a rubber pinch-roller ensures that the tape-speed does not fluctuate. The remaining two motors are called the 'Torque Motors'. These apply equal and opposite torques to the input and take-up of reels during recording and play-back functions, which maintain the tape's tension. During fast winding operations the pinch roller is disengaged and the take-up reel motor is supplied with a higher voltage than the supply motor. In certain cheaper models, a single motor can do all these functions. In such a case; the motor drives the capstan directly, while the supply and take-up reels are loosely coupled to the capstan motor with slipping belts or clutches. There are also tape-recorder variants with two motors, in which one motor is used for rewinding only.

6.10 Magnetic Tape Recording

An electric current flowing in the coils of the tape-recording head produces a fluctuating magnetic-field. This, in turn, causes the magnetic material, coated on the tape, and moving past the recording-head, to align in a manner proportional to the original signal of sound. The recorded signal can be reproduced by running the tape back across the tape head, when the reverse process occurs. The magnetic imprint on the tape induces a small current in the read-head, which approximates the original signal and it is then amplified for playback by a loud-speaker.

Just like gramophone records, tape-recording is also an analog-process. Since its first introduction, during mid-1950s, it has been progressively improved and developed by the manufacturing companies. This has resulted in an improvement of sound quality, convenience, and versatility of the machine.

Two-track, and multi-track heads have also been designed, which permits, discrete recording and playback of individual sound sources, by stereophonic channels, or different microphones during live recording.

6.11 Designer's Tape Recorder

These are more sophisticated machines, which have separate heads for recording and playback. Additionally there is also a separate head for erasing. This can be used in monitoring of the recorded signal very quickly, when the recording is being done. By mixing up of the playback signal, into the recorded input creates a primitive echo generator. Dynamic range compression during recording and expansion during playback expands the available dynamic range and improvement of the signal-to-noise ratio.

Dbx and Dolby Laboratories have introduced add-on products in this area, which were originally meant for studio use only. Later on other versions of such machine were also put in the consumer markets, for personal use.

Currently, the tape-recording of sound is the most commonly used practice, even though the technology of 'digital audio recording' is fast emerging as its competitor.

6.12 Emergence of Digital-Audio and Compact-Disc Recording

Digital audio recording and sound reproduction involve the use of pulse-code modulation and digital signals. Hence the 'Digital Audio Systems' includes, 'analog-to-digital' conversion (ADC), as also 'digital-to-analog' reconversion (DAC) systems. Additionally it also includes digital storage, processing and transmission components. The main advantage and usefulness of digital audio-recording is its ability to store, retrieve and transmit signals without any loss in the quality of sound.

Since late 1980s, digital audio recording has gained a much larger share in market of sound recording, when compared to that of magnetic tape-recording. However, recording on magnetic tapes is likely to continue in even in the 21st century and it has a very good market for the audio recorded goods.

There has been some recent report about superiority and fineness in sound recording on the conventional disc-record, when compared to that on magnetic-tape or CD's. To certain extent, this has resulted in reemergence of gramophone records in many developed countries. Since the year 2008, vinyl gramophone-records have regained their popularity, in USA and Europe. According to one report, nearly 2.9 million units of these disc records were manufactured and sold during the year 2008. This has been the highest ever since the year 1998.

Gramophone recording are particularly used by the 'DJs' and 'Audiophiles' for many types of music. As of the year 2011 in USA, vinyl records continue to be the source music used for distribution of independent and alternative music artists. More mainstream pop music releases, tend to be mostly sold in compact disc or other digital formats, but vinyl-discs have still been released in certain instances.

It may therefore be concluded that market for magnetic tapes and digital CDs is likely to exist along with the old vinyl-discs, and music-systems are now available, which can play either or all of these.

Fig.-1, Edison's Cylinder Gramophone of 1899
Photo, Norman Bruderhofer, article on gramophone,
http://en.wikipedia.org/

Fig.-2, A 1936 Model of 78 rpm, Mechanical, Flat-Disc Gramophone
Photo, Luekk, article on gramophone, http://en.wikipedia.org completely
and only the digital CD's are now used in PC's

Chapter-7

❖

Magnetic Tapes for Audio and Video Recording

7.1 Introduction

Magnetic tapes are devices for analog recording and storing of data, which are based on electronic signals, and these can be recalled, when desired. The data most commonly recorded on magnetic tapes are those based on audio and video performances. Earlier to the introduction of magnetic tapes, the recording of sound was exclusively done on gramophone records. Similarly video recording was only done by photo-filming on plastic films, from which they were copied on paper. A photo, filmed on a plastic film in the form of a 'slide' could be projected (See also **Chapter-3, 4**) on a screen. If continuously filmed on a moving object, it could also be projected on screen as a motion picture. Thus, on one hand the invention of magnetic tape has revolutionized recording and broadcasting of audio and video, while on other hand it has made data storage and processing much easier.

Magnetic tape recording is an analog system. It has been a key technology, in early development of computers, allowing unparalleled amount of data to be created, stored for long periods, and copied when desired and to be accessed quickly.

Due to several early refinements in the technology of sound recording and reproducing by the magnetic tapes, the system has become one of the best quality of analog sound recording medium available. Even today, these are considered to be one of the best systems of recording. Since the first decade of twenty-first century, the digital system of sound and video recording has to certain extant emerged as an alternative to magnetic tapes in certain areas, particularly for sound and video recording, but has not completely replaced magnetic tape for sound and video recording. Any many specific areas, use of magnetic tapes has continued, unchallenged.

7.2 History of Magnetic Tape

Early developments on magnetic tapes took place in Germany, but due to the escalating political tensions, and the outbreak of the Second World-War in the year 1939, these developments were largely kept secret by Germans. The Allies came to know about this development from their monitoring of Nazi Radio-Broadcasts. They concluded that the Germans have developed a completely new sound recording technology. The nature of this technology remained undiscovered, until the Allies victory, and capturing of German recording equipment, at the end of war, i.e. the year 1946. It was only after the war that some American technologists, particularly Jack Mullin, John Herbert Orr, and Richard H. Ranger were able to bring this technology, out of Germany. In the USA they further developed it into a commercially viable form.

The basic idea of magnetic tape recording was conceived as early as the year 1877, by an American engineer, Oberlin Smith, but its first application was demonstrated only in the year 1898, by a Danish engineer, Vladimir Paulsen. Analog based magnetic recording on tape, as it was conceived by Oberlin Smith, now uses a long strip of thin plastic strip, coated with a magnetizable medium, which is an oxides of iron. The tape moves, with a constant speed, past a recording head. An electrical signal, which is analogous to the sound to be recorded, is fed into the recording head, and it induces a pattern of magnetization similar to the signal, on the tape.

7.3 Invention and Basic Technology of Magnetic tape

Narrow plastic strips, on coating with a magnetizable substance, i.e., ferric oxide, on them are made into magnetic tapes. In Germany, during the year 1928, Fritz Pfleumer first invented magnetic tapes for recording sound. This was based on an earlier invention of 'magnetic wire recording' invented by Valdemar Poulsen in the year 1898. Fritz Pfleumer used ferric oxide (Fe_2O_3) powder coated on long strips of paper. This invention was further exploited by the technologists of German Electronics Company (AEG). The company also manufactured a magnetic tape recording machines, in the year 1933. Soon after this, the BASF-Corporation of Germany started manufacturing of magnetic tapes.

A ring shaped tape-recording head was developed by Eduard Schuller, for tape recording machine. Earlier designed recording heads, were needle shaped these were discontinued due to their tendency to shred off the moving tape.

An important discovery made during this period was the technique of 'AC Biasing'. This new technology improved the quality of the recorded audio signal by increasing the effective linearity of the recording medium. For playing back the recorded sound, the play-back head could then pick up the changes in magnetic field, from the tape and convert these into an electronic signals and ultimately into an audio signals, with the help of suitable transducer, i.e. a loud speaker.

7.4 Earlier Magnetic Recording on Steel Tape

A thin steel wire was initially used for magnetic recording. Since the wire used to cut and damage the recording head, it was later on replaced by a thin steel tape. Such a recorder, which was called, the 'Blattnerphone', was designed in the year 1930 by the 'Ludwig Blattner-Picture Corporation'. This was followed by another engineer, named Clarence N. Hickman, of the Bell Telephone Laboratories in USA, who made a prototype of a steel-tape recorder based, telephone-answering machine.

On the Christmas Day of the year 1932, the British-Broadcasting-Company (BBC) made its first use of a tape recorder for their broadcasts. The device used by the BBC, was a 'Marconi-Stille Recorder', with a 3.0mm wide and 0.08mm thick, razor-steel tape. For reproducing higher audio frequencies, the speed of the tape, which passed over a recording-cum-playing head, was kept at 150cm per second.

This meant that the length of a tape, for a program of half an hour had to be nearly 3 kilo-meter in length, and which would weigh nearly 25 kg. For safety reasons such machines were only operated in a locked room, using a remote control devise. Due to the tape's speed, springiness and razor-like sharp edges, if the tape broke while in operation, it could unspool, and fly off to cause serious injury to an operator. Ultimately this could also lead to massive data loss and poor audio quality.

By the middle of the year 1930, the C. Lorenz Company in Germany had developed a steel tape recorder that became quite popular with telephone companies in Europe, and particularly by the German radio networks.

In the year 1938, a German designer and engineer, Begun, left Germany and joined the 'Brush Development Company' (BDC) in USA, where he continued to work on similar developments, related to magnetic tapes and tape-recorders. In his capacity as a development engineer, he got little attention of company, which employed him, until the late 1940's. The BDC, in the year 1946, released its first commercial tape recorder. This machine was called, 'Soundmirror BK 401'. Several other models of the machine were also released in the following years.

It has been mentioned earlier that the tapes were initially made of paper, which were coated with black oxide of iron. Minnesota Mining & Manufacturing Company of USA, in the year 1948, replaced paper tapes by plastic tapes.

7.5 German Developments on Plastic Coated Magnetic Tapes

Magnetic tape recording, as we know of it today, was mainly developed in Germany during the year 1930, by the BASF Company, which was a part of the chemical giant, IG. Farben company; in cooperation with certain other companies and it was based on Fritz Pfleumer's early invention of paper tape with iron oxide powder lacquered to it.

Engineers at the AEG-corporation in Germany, also designed the world's first practical magnetic tape recorder, and named it 'K1'. This machine was demonstrated it in the year 1935. Eduard Schüller, an engineer at AEG built the recorders and developed a ring-shaped recording and playback head, and replaced it in place of the earlier used, needle shaped head. Friedrich Matthias of 'IG Farben and BASF' has been credited to have developed the recording tape, including iron-oxide powder, a suitable binder for it and a backing material i.e., the plastic-tape. Later on, Walter Weber, working for 'Hans Joachim von Braunmühl' at the RRG, discovered the AC-biasing technology for improved the quality of recorded sound.

7.6 The World comes to knows about the German Invention

During the Second World-War (1939-1946), it was noticed by the Allies that certain German officials were, almost simultaneously making radio broadcasts from different time zones in Europe. An investigator, namely Richard H. Ranger then concluded that these broadcasts had to be transcriptions from 'recorded speeches'. Even the audio-quality of these broadcasts was very much similar to that of a live-broadcast. It was further observed that the duration of these broadcasts, was much longer than what was possible with then-known 78-rpm gramophone discs-records. During the final stages of the Second World-War in Europe, i.e. the year 1945, the Allied forces captured a number of German 'Magnetophon-recorders' or the tape recorders from the Radio-Station at Luxembourg. These recorders incorporated all the key technological features of modern, analog magnetic tape-recording, and these formed the basis for future developments in this field in USA.

7.7 Developments related to Tape Recorder in USA

Development of magnetic tape recorders, in the USA, was carried in the late 1940s and early 1950s, by the 'Brush Development Company' and its licensee-company, 'Ampex'. An equally important development of the magnetic tape media was led by the engineers of the 'Minnesota Mining and Manufacturing Corporation, which was also known as the 3M-Corp.

An American audio engineer, John T. Mullin and a music entertainer named Bing Crosby were the two key players, who developed and commercialized magnetic tapes in USA. At that time Mullin was serving in the 'U.S. Army Signal Corps', and during the final days of the Second World-War, he was posted in Europe at Paris. His unit was assigned to find out everything, they could, about the system of German-Radio and their electronic technology.

The scope this investigation included the claims that at that time the Germans were experimenting with a high-energy technology, which was based on 'directed radio-beams'. These radio-beams were meant for disabling the electrical systems of an aircraft. Mullin's unit soon found a huge collection of magnetic-dictating machines, which were of rather a poor quality. He visited a studio at Bad Nauheim, near Frankfurt, and while investigating radio-beam, he got certain rumors, which was considered to be the real prize, i.e. the very-basics of tape-recording technology.

7.8 Innovative Uses of Tape Recording

In the year 1948, the first commercial tape recorder, the 'Ampex 200' model, was launched in USA. An American musician, cum inventor, Les Paul, invented the first 'multi-track tape recorder'. This multiple-track on the recording tape was another technical advancement in the sound recording and playing system **(fig.-1)**.

Magnetic tape could also make it possible, to record the sound which was totally created by electronic means. This created a the way of bold, sonic experiments, pioneered by the 'Musique Concrète' School' and 'Avant-Garde Composers' such as Karlheinz Stockhausen. Further, it led to the innovative development of 'Pop-Music Studio Recordings' by artists, groups such as those of Frank Zappa, Beatles and the Beach-Boys.

Tape recording has also enabled the radio and TV broad-casting industry to pre-record their commercial advertising, and helped in creation and duplication of the complex, high-fidelity, and long-duration recordings of programs.

An important limitation of magnetic tapes still remained. After a long duration of time, these tapes underwent a type of deterioration, which has been termed as the 'sticky-shed syndrome'. This is caused by moisture absorption into the binder material of the tape, which can, ultimately render the tape unsuitable for use.

Till about around the year 1920, gramophone record was the only known method for recording sound. Introduction of magnetic tape for audio recording, which started around the year 1930, completely revolutionized radio broadcast and music recording systems. Tape recording has given artists, singers and producers a power to record and re-record an audio, without any loss in its quality and to edit it, as they like and rearrange the recordings, whenever it is necessary.

7.9 Video Recording on Magnetic Tape

Many inventors also saw potential of using magnetic tape recording of video for television show. Hence, the sound recording on magnetic tapes, was soon followed its use in video-recording and playing. The electronic video-signals are somewhat similar to the audio-signals. The major difference between these two being, that the video-signals use more band-width, when compared to audio-signals. Hence the normal magnetic audio tape-recorders could not capture a video-signal.

Very soon, some inventors of audio tapes started work to find a suitable solution for this problem. Thus, Jack Mullin, who worked for Bing Crosby Co. and the BBC Corp., collaborated to create a working system, which involved a tape moving across, a fixed tape-head at a very high speed; but this did not work well. Later on Charles Ginsburg and his team, at the Ampex Co., could make a real breakthrough. They used a spinning recording-head, with which at a normal tape speed, he could achieve a very high head-to-tape speed. This could record and reproduce the high bandwidth of video-signals. The Ampex-system was called 'Quadruplex' and it used 2-inch (51 mm) wide tape, mounted on reels like audio tape, which recorded the signal in what is now called a 'transverse scan'.

In a later improvement in video-tape recording, by the Sony-Corporation of Japan, it developed a helical scan technology and enclosed the tape reels in an 'easy-to-handle' cartridge. Nearly all modern video-tape recording systems use helical scan and cartridges. Video-cassette recorders are now very common in homes as well as in television production facilities in industries.

With the introduction of digital-video and computerized video processing, optical disc media and digital video-recorders can now, perform the same role as magnetic video-tape. These devices also offer improvements like random access to any scene in the recording and 'live' time shifting and have replaced videotape in many situations.

7.10 Magnetic Tape for Data Storage

In all the cases of tape-formats, a 'tape drive', which is also called a 'transport' or a 'deck', uses motors to wind the tape from one reel to another. During passing past the tape heads, it can read, write or erase, as the tape moves.

In the year 1951, for the first time, the magnetic tape was used for recording computer-data on a machine, which was 'Eckert-Mauchly UNIVAC-I'. The recording medium was a strip of a 12.65 mm or half an inch wide metal, consisting of nickel-plated bronze called 'Vicalloy'. Recording density was 128 characters per inch i.e.198 micro-meter per character on eight tracks.

Early IBM tape drives were 'on the floor-standing drives' that used vacuum columns to physically buffer long U-shaped loops of tape. Two tape-reels were visibly fed through the columns, intermittently spinning the reels in rapid, unsynchronized bursts, resulting in a visually striking action. Stock shots of such vacuum-column tape drives in motion were widely used to represent 'the computer' in movies and television.

Even now, magnetic tape remains a viable alternative to digital compact-disk (CD) in some situations due to its lower cost per bit. This is a big advantage when dealing with large amounts of data. Though the real recording density of tape is lower when compared to compact-disk, the available surface area on a tape is far greater (Fig.1). The highest capacity tape media are generally on the same order as the largest available disk drives, which have achieved about 5 TB in the year 2011. Historically, the tape has offered enough advantage in cost over compact-disk storage of data, to make it a viable product, particularly for backup, where media removability is necessary.

In case of computers, the introduction of digital-technology, has very much reduced their size, which have now become table-models (PC) and even 'Lap-top'. Earlier models of PC's used magnetic tape, in the form of 'floppy', which are now phased out completely and only the digital CD's are now used in PC's

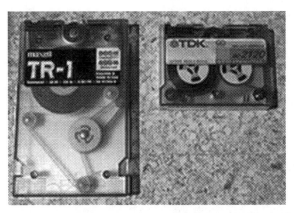

Fig.-1 Quarter inch magnetic tape cartridges and Tape-Recorder..
Source, Photo, Wikimedia Commons, http://en.wikipedia.org/

Chapter-8

⋘✧⋙

Conventional Pendulum Clock, *verses* Quartz Clock

8.1 Introduction: Pendulum Clock

A pendulum, which generally consists of a swinging bob of a circular metal piece, has long been used in a conventional, mechanical clock for its timekeeping mechanism. The pendulum of a clock is a mechanical, resonating device, which swings, back and forth, in a precise interval of time. The time interval of a pendulum depends upon the effective length of a pendulum. Ever since the invention of a pendulum by Christian Huygens in the year 1656, pendulum clocks have been acting as the precise time-keepers throughout the world.

It is necessary to keep a pendulum clock in a completely stationary position, during its operation. This is because any acceleration, resulting due to its motion, can cause change in its swing-period. This, in turn can ultimately result in inaccuracies in timekeeping. This has been the reason for not using pendulum mechanism in portable timepieces. Bulkiness of a pendulum completely rules out its use in wrist and pocket watches and table time-pieces.

8.2 Pendulum as a Resonate Device

In the year 1606, the famous astronomer cum scientist, Galileo Galilees discovered a property of pendulum, which makes it useful as a timekeeper. A pendulum is a isochronic devise, which means that the 'swing period' of a pendulum is independent of the 'swing-amplitude'. Swing period means the time it takes to complete one complete swing and return to one of the extreme displaced position on any particular direction. Swing-amplitude is the maximum distance to which it is displaced from the mean position, when it is stationary. This is true only when swing-amplitude is kept small.

In history it has been described well. Galileo, during a weekly service in a church in Italy, observed that a swinging, decorative chandelier (Hindi, *Fanus*), in the church did not change its period of a swing, with a small change in its amplitude. He estimated the constancy of swing-period, by counting a number of swings, against a fixed number of his heart-beats (pulses). Based on this observation, Galileo conceived the idea of a pendulum clock in the year 1637, and assigned the work of its fabrication to his son, in the year 1649. His son, however, did not live long enough to finish, designing of a pendulum clock.

A Dutch scientist, Christian Huygens, has been credited for the invention of pendulum clock, in the year 1656, which he patented a year later. Huygens was inspired by Galileo's observation that the swing period of a pendulum is independent of swing amplitude, when it is kept small. Huygens assigned a contract for construction of a clock designed by him, to a clockmaker, named, Solomon Costar, who built the first pendulum clock for him. Even at that time, the introduction of the pendulum in a clock, as a harmonic oscillator for timekeeping, increased the accuracy of clocks to about 15 seconds per day. This was a great achievement.

The eighteenth and nineteenth centuries have been a period of 'horological' (time-keeping) innovations. During this period, the invention of the pendulum brought many improvements in clock making. Earlier to this, i.e., in the year 1675, Richard Towneley invented a mechanism, called 'Deadbeat Escapement', which kept the swing amplitude of a clock-pendulum small. This invention was further popularized by a mechanic cum technologist, George Graham in the year 1715. In his precision regulator clocks, Graham made use of a device, known as 'Anchor Escapement', which is even now used in most of the modern pendulum clocks (**Fig.-1**).

Early clocks had wide pendulum swings of up to 100 degrees. 'Winged Lantern Clock', was an early type of pendulum clock, made by a British clockmaker, named Edward East in the year 1657, in which 'wings' were added to accommodate large pendulum swings. Later on it was shown by Huygens that wide swings made the pendulum's swing-period more variable. This made clock-makers to realize that pendulums with small swings (<10 degrees) are isochronous and this was achieved by invention of 'Anchor Escapement'.

Very tall and narrow clocks, which are also known as 'Grand-Father Clock' **(Fig.-2)**, were first built by William Clement in 1680. These clocks had a 'Second's Pendulum', i.e. swing period of one second. Second's pendulum also came to be known as "Royal Pendulum'. The length of a second's pendulum, having a swing period of one second is nearly one meter. Some early clocks had only 'hours hand', and it was later on, when accuracy increased, that the minute and even second's hand, were added on clock faces.

8.3 Temperature Effect in Pendulum Clocks

When it was observed that most of the pendulum clocks slowed down during summer, it was realized that thermal expansion and contraction, of the effective length of pendulum rod, with a seasonal changes in temperature, was a source of this error. To some extent this problem was solved, by the invention of temperature-compensated pendulums.

One such device was a 'mercury pendulum', designed by George Graham, in the year 1721. The mercury pendulum was made of a hollow metallic bob, which was partially filled with mercury. Any change in the effective length of the pendulum, due to temperature change, was compensated due the pendulum rod and mercury in its bob, expending in opposite directions.

Yet another device was a 'grid-iron pendulum', which was designed by John Harrison, in 1726. In case of a grid-iron pendulum, a grid of iron rods was so constructed that the rods in the grid, expended in opposite direction and kept the effective pendulum length constant.

Due to these improvements, by the mid-eighteenth century, certain precision pendulum clocks achieved accuracy of the order of a few seconds per week. During the nineteenth century, pendulum clocks were still handmade by highly skilled craftsmen and this made them very expensive. Ornamental pendulum clocks of this period were indicator their value, as well as status symbols of the wealthy persons.

'Industrial-Revolution', starting in early seventeenth century, resulted in a better and organized life-style, even for a common person, and during this period. This made a 'home pendulum clock' became his time-keeper in a family. Very special 'Regulator clocks' of that period were highly accurate pendulum clocks during that time, and these were installed in public places, where these were used to schedule work, and to set time in other clocks. Earlier to this, i.e. at the beginning in the nineteenth century, astronomical time regulators in observatories served as primary standards for national time services.

Manufacturing of hand-made clocks, during that period, provided living to a large number of highly skilled craftsmen. These artisans specialized in manufacture of decorative clocks, and did repair and maintenance of clocks, timepieces and watches. It was common to see many exclusive clock and watch repair shops in towns in India and elsewhere manned by highly skilled persons Pendulum clocks remained the world's standard for accurate time-keeping for more than 250 years, until the first invention of the quartz clocks in the year 1927. Still it took a long time for quartz based time keeping machines to replace pendulum clocks.

8.4 Quartz-Clock and Watches

In contrast to pendulum clocks, a quartz clock uses a quartz crystal as an electronic-oscillator (resonator) for timekeeping. Naturally occurring, crystalline quartz is a highly pure variety of silica (silicon dioxide, SiO_2) which forms large, glass like transparent crystals. Silica is an abundant and widely distributed mineral. Quartz has the property of having an extremely low thermal expansion, and thus it retains its size, when there is temperature change from summer to winter: i.e. it does not change appreciably in dimensions, with change in temperature. For this reason, fused quartz is also used for making certain laboratory wares, e.g., crucibles for high temperature use.

Quartz, like many other materials can be machined into thin plates, which will resonate. Quartz crystal can be directly flexed by an electric signal. Resonance frequency of a quartz plate of a definite size will not change significantly with change in temperature, and hence a quartz clock will remain more accurate at ambient temperature. A quartz crystal, as an oscillator, creates a very precise frequency signal. Hence quartz clocks are far more accurate than most good mechanical (pendulum) clocks. A digital logic counts the cycles of occupations and provides a numerical time display, in units of hours, minutes, and seconds. Today, quartz based timekeepers are the world's most widely used timekeeping machines.

8.5 Mechanism of Quartz Clock

A properly cut and mounted quartz crystal oscillates, on applying an electrical AC (signal) to it through a pair of electrodes. Such oscillation, in turn, causes a voltage change across the crystal, which can be detected by a suitable electronic circuit.

This property of quartz to generate an electric current upon the application of mechanical pressure, and *vice versa*, is called, a direct and a reverse, 'Piezoelectric-Effect'.

Resonator quartz crystal, used in a clock acts like a small tuning-fork, vibrating at a fixed frequency of 32,768 Hz. In a simple chain of digital divided by 2, this stage can be derived as 1Hz signal, which is required to drive the second's hand of a quartz clock. The frequency at which the crystal oscillates is dependent on its shape, size, and the positions at which electrodes are placed on it. If the crystal is accurately shaped (4.0 X 4.0 mm and 0.3 mm thick) and positioned, it will oscillate at a frequency of 32,768 Hz. This frequency is fifteenth power of two (2^{15}), which can easily be counted using a 15-bit binary digital counter. Once the circuit supplying power to the crystal counts this number of oscillations, it increases the recorded time by one second.

8.6 Visual Display and Choice of Battery

Quartz clocks and watches are made to have either a digital or an analog display of time, and these are generally battery operated. Use of large size quartz-resonator of lower frequency will result in more current use, and this shall lower battery life. Currently by using integrated circuits and metal oxide based semiconductor, a battery life of 12 months for a coin-cell is easily achieved in quartz wrist-watches. In quartz clock, a dry battery (pencil cell type) also has a 12 months life. A single coin cell, driving a mechanical stepper motor for indexing the second's hand, of a quartz analog watch, can also have about 12 months life. Quartz clocks **(Fig.-3)** and watches are also available with digital display, based on liquid crystal, in which the earlier used light emitting diode display system has now been phased out, to reduce battery consumption.

8.7 Accuracy of Quartz Clock

The relative stability of a quartz-resonator and its driving circuit is much better, in a quartz clock, than its absolute accuracy. Standard-quality resonators of this type, carry a warranty to have a long-term accuracy of about six parts per million at 30°C. This means that a typical quartz clock or wristwatch will gain or lose less than half a second per day at body temperature.

If a quartz wristwatch is rated by measuring its timekeeping characteristics against atomic clock, to determine how much time the watch gains or losses per day, then the corrected time will easily be accurate to within 10 seconds per year.

8.8 Chronometers

The term 'chronometer' is used for an extremely accurate clock or a watch, which can serve as a timekeeper for celebration of other clocks and watches. Earlier to the introduction of quartz chronometers, specially designed and very accurate pendulum clocks were used as time standards.

Today quartz wristwatches are in high demand, because they are more accurate than their mechanical counterparts. They do not need regular winding and their maintenance is also easy. Sunlight powered and motion-powered quartz watches are also made, which represent two innovative types of watches. Light-powered quartz watches incorporate a solar cell, which transforms solar light into electricity, while in a motion powered quartz watch, there is a tiny rotor spinning in response to motion of the wrist, and generate electricity.

8.9 The Future Prospects and Conclusion

Throughout the world, the pendulum clocks are still integral parts of monumental clock towers, where these are likely to continue to operate their specially fabricated, old pendulum clocks. For a common man the end of conventional pendulum clock era has already arrived. Most clock manufacturers, throughout the world, have almost stopped pendulum clock manufacture. The basic units i.e. the machine portion for making a quartz clock or watch is now being made on large commercial scale by countries like China. The machine portions are imported into different countries, such as India, where a variety of attractive and decorative clock faces, are fixed on them, as desired. Such clocks and watches are very cheap. However, man's association with pendulum as a part of a clock has been so strong that even quartz clock makers, are now adding pendulum to their clocks, more for a show.

Fig-1. A common Pendulum Clock, which uses 'Anchor Escapement' to restrict pendulum

Photo, Av Kumar, article on Pendulum Clock, http://en.wikipedia.org/

Fig-2. Tall and narrow (Grand-Father) clock with second's pendulum

Photo, Luekk, article on Pendulum Clock, http://en.wikipedia.org/

Fig-5. A Quartz Clock

Photo, Cannon PowerShot, article on Pendulum Clock,
http://en.wikipedia.org/

Chapter-9

Telephone for Telecommunication of Voice

9.1 Introduction

The word telephone, is derived from two Greek words, tēle (τῆλε), which means far-distant, and phōnē (φωνή), meaning sound. A simplified and shortened term, often used for telephone is 'phone'. The terms telephone or simply phone are colloquially referred to telecommunication devices, while the term telephoning is used for transmitting and receiving, human conversation. The basic function of a telephone is considered to be a point-to-point communication, between persons separated by a distance. The system allows distant people to talk, just as if they were sitting across a table. Telephone has now become common appliances in the developed as well as under-developed countries. Presently a telephone is considered to be an indispensable facility for any business establishment, government establishments, as well as offices and a private household. The word 'telephone' has now been adopted in many languages and in India it is also called *'door-bhas'* (in Hindi) meaning 'distance talk'.

9.2 Alexander Graham Bell's Invention and Patenting of Telephone

There have been a large number claims and counterclaims as to the first inventor of telephone. Similarly many claims for improvements of telephone and services related to it have never been settled. Thus the credit for the invention of the electric or electronic telephone devise has frequently been disputed by different persons. Even new controversies over this issue have arisen from time to time. Even a large number of lawsuits could not settled down for patent claims, related to telephone, which were made by several individual persons and commercial competitors.

In fact like many other industrial inventions, e.g., radio, television, light bulb, and computer, there have been several inventors, who did pioneering, experimental work, on sound transmission on a wire. Due to the work attributed to many inventors, wireless transmission has developed, to the extant, we see it today.

Thus, Innocenzo Manzetti, Antonio Meucci, Johann Philipp Reis, Elisha Gray, Alexander Graham Bell, and Thomas Alva Edison, are among several persons, who have been credited with pioneering work on development of telephone.

However, it is an undisputed fact that it was Alexander Graham Bell, who was the first person to be granted a patent for electric telephone by the Patent and Trademark Office (USPTO) of the United States of America's, in the month of March, 1876. The first patent, by Alexander Graham Bell can be considered as the 'master patent' on the telephone, from which all other patents for electric telephoning devices, have originated. It has already been mentioned earlier, that there were many individuals and industries, which have claimed patents for invention of telephone. Ultimately it was the 'Bell Telephone Company' and Edison's patents, which were victorious and commercially decisive.

When a workable telephone instrument was ultimately designed, there arose a need to design a system to interconnect different telephones. A Hungarian engineer named Tivadar Puskás, in the year 1876, designed a switch board for interconnecting different telephones. This invention, which followed the original invention of telephone instrument, has made the practical use of telephone possible, by allowing the formation of large telephone-exchanges, and eventually the world-wide telephone networks, which we have today.

9.3 Components of a Simple Telephone

A simple (somewhat old model now) telephone instrument is shown in **Fig.-1**. It comprises of a hand-set, which consists of a microphone, to speak into at one end by a (first) person. At the other end of this hand-set, there is an earphone, which reproduces the voice of another (second) person, to whom the first person is talking. Additionally, there is a telephone ring system, which makes a ringing sound to alert the person, when an 'out-call' is coming. Initially there were telephone-exchanges in many under-developed countries, including in India, where a connection between the persons desiring to talk was made manually, and on a request to a well trained telephone-operator working at the exchange. Nearly all over the world now, completely automatic telephone exchange systems (**Fig.-2)** are in operation.

A keypad is provided, with each instrument to enter the number of the telephone being called. In some-what older telephones, **(Fig.-1)** there used to be a manual, dialing system, based on rotating numbers? These have now been replaced by press-buttons system. The handset having both a microphone and an earphone built at its two ends is held, near the mouth and an ear when talking. The keypad, used to be the part of telephone base, but in newer instruments, it may be a part of the portable handset.

A landline telephone is connected by a pair of wires to the telephone-exchange i.e. the offices of telephone network, extending over cities, countries and even the whole world. In contrast to a land-line telephone, a mobile-phone or cell-phone is portable and not connected by any wire-line. It communicates via the vast land-line telephone network, or by the wireless-radio, using a base station connected by wire to the telephone network. Earlier a mobile telephone could only be used within a limited range of the base station, but now they can be used to talk anywhere in the world.

9.4 Principles of Telecommunication

A person having a telephone is allotted a specific telephone number. This person has to dial the telephone number of another person, to whom he wants to talk. The person at the other end receives a ring or a signal to indicate that someone is calling him. When he lifts the hand-set a connection is automatically made between the two persons. As a person talks into the telephone microphone, it creates electric impulses, which are carried, on line or by radio-waves, as the case may be, to the telephone of another person. This is possible in both ways, and hence the two persons can confer as if they were sitting together.

An important limitation of land-line telephone has been that it has to be located at fixed place, which could be a home, an office or a public telephone-booth. Public telephone-booths were generally located at certain common places e.g. at the railway and bus stations and airports, and many other busy places in a city. These were particularly useful for persons, away from home and not in possession of a mobile-phone. In most of the cases, the automatic public telephone booths are coin operated and these could be activated by depositing exact coins. Alternatively these could also be operated through a telephone operator, who can be asked to charge for a call to a specific telephone number of the receiver (collect-call).

9.5 Mobile-Telephone, Cell-Phone or the Hand Phone

The latest development in telephone technology has been the 'mobile-telephone'. A mobile-phone **(Fig.-3)** is also called a cell-phone or a hand-phone. It is an electronic device, which is

105

used to make mobile telephone calls i.e. even when a person is moving or travelling. Mobile telephone now covers a wide range of geographic area, i.e. almost all the habituated world. It is now being served by many public cells, and thus allows service to be mobile at the time of making a call. By contrast, a cordless telephone can be used only within the range of a single, private base station, for example within a home or an office.

A mobile phone can make and receive telephone calls, to and from any public telephone network, which includes other mobile and land-line phones across the world. This is done by connecting it to a cellular network provided by a mobile network operator.

9.6 Multiple uses of a Mobile Phone

In addition to telephoning, the manufacturers of modern mobile phones also provide a wide variety of many other services in these handy instruments. These include, text—messaging, SMS, e-mail, internet access, short-range wireless communications, business applications, gaming and color-photography. Mobile phones, which provide these more general and computing capabilities, are commonly referred to as 'Smartphone's.

The first hand-held mobile phone was constructed and its use demonstrated by Dr. Martin Cooper, of Motorola Corp. in the year 1973, using a handset, which weighed about 2 kg. In the year 1983, the model 'DynaTAC 8000x' was the first mobile phone to be made commercially available. Pocket held, and very small mobile phones are now manufactured (**Fig.-3**). During last twenty years (1990 – 2010), worldwide mobile phone users have grown from 12.4 million to more than 4.6 billion, which indicates its usefulness and popularity.

Fig.-1. An Olivetti Rotary-dial Telephone (1940)

Photo, Kornelle & H.Hafele, Article on telephone,

Article http://en.wikipedia.org/

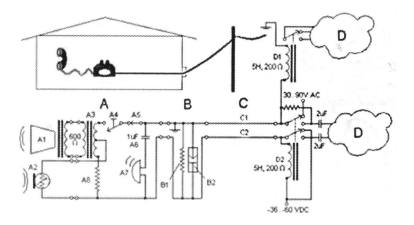

Fig.-2. Typical scheme of a landline telephone installation.

Photo, Kornelle & H.Hafele, Article on telephone,

Article http://en.wikipedia.org/

Fig.-3 A Typical Mobil-Phone

Photo, Kristoferb, Article on telephone, http://en.wikipedia.org/

Chapter-10

Decline of Conventional Telegraphy and Emergence of E-mail and Fax-Services

10.1 Introduction: Telegraphy, Indian Post & Telegraph Department

Post & Telegraph Department has been one of oldest and efficient departments of the Government on India. It has been functioning much earlier to India's independence year of 1947. An efficient system of collecting and distributing, letters and telegrams, even to the remotest places in the country, which were not even connected by rail or road, has been in practice in India, since mid-nineteenth century. This, postal-network covered all the British-ruled regions of pre-independence India, as well as all the princely states of that time. In some of the princely-states there were additional postal systems, catering within the state only.

In India, Post & Telegraphy services have always remained with the central government. Letters, by regular post has to go by train or road transport and takes time. If it gets delayed, the very purpose of sending an urgent message is defeated. Sometimes it happens that a massage is not quickly delivered to the intended person, even after the concerned post-office has received it. Hence for a fast communication of an urgent massage telegram service becomes necessary.

10.2 Telegraphy and Rail-Services

A large and expending railway network (owned by several private companies) existed in India, since the late nineteenth century. The railway companies required an efficient system of connecting different railway stations for communicating, the arrival and departure train timings. Invariably a telegraph line was installed along with the rail-line and it was a common facility, being used by the 'Railways' and the 'Post & Telegraph Department' at that time. In many small places (stations) the work of sending and receiving private telegrams was also done by the trained railway staff, and there used to a post of *'Tar-Babu'* i.e. a person trained in telegraphic work, beside other station staff. The postal department paid them extra for the work. In fact there was an enviable coordination, between different departments, and sometimes it was common for a village schoolteacher, to work part-time, for postal department. He did the selling of postal material, sorting of post (Hindi, *Dak*) and its distribution by a post-man (*Dakia*), who was the sole employee of postal department and, who sometimes covered the distribution of *dak*, in several villages on foot.

It will be of interest to know about the *dak* distribution system in the erstwhile Udaipur or the Mewar state, where a *'Brahamin'*, was particularly employed for distribution of *dak* in villages within the state and this was called *'Brahamini Dak'*. High-way robbery was common at that time, but even the 'God-Fearing Robbers' never attacked a *Brahamin* and the *Dak* was delivered safely.

10.3 Defining Telegraphy

The word 'telegraphy' is derived from two Greek words, i.e., '*tele (τηλ)ε)*', meaning distance and '*graphein*' (*γραψειν*), which means to write. In other words it means, 'A long-distance transmission of a written messages' without any physical transport, such as that of a letter. There are certain other terms related to telegraphy e.g. 'Optical-Telegraphy' and 'Radio-Telegraphy', which means 'observing from a distance', and 'wireless telegraphic transmission' of a messages using radio-waves', respectively. Currently, the expression, telegraphy, includes the more recently developed forms of data transmission such as fax, e-mail, and general computer network.

10.4 Invention of Telegraphy and Morse-Code

The idea of sending massages by electrically produced sound signals, was conceived in the year 1832, by Samuel F. B. Morse, who was a painter by profession, and founder of the 'National Academy of Design' in USA. Only two sound signals, were necessary and these were designated 'Dot' (•) and Dash (-). Typical representation of all the English alphabets and numbers in Morse-Code is shown in **Fig.-1.** By the year 1836, Morse produced a working, model of a electromagnetic telegraph machine. The first telegraph massage, conveyed by him, which was based on Morse-code, read as,-

"What hath God wrought?"

This massage was sent from Washington to Baltimore, in USA.

An interesting similarity can now be seen between Morse-cod and the modern, digital communication technology, both of which are binary systems. The later also uses only two digits (0 and 1). Not only all the alphabets and numbers, in the form of well punctuated words and massages, but any other electronic signals, which may be based on sound, color and light intensity etc. can be expressed digitally and electronically. Had electronics, been well developed, during the period, when Morse-code was invented, then it could have become as good as today's, digital technology!

Initially the telegraph operators, used to copy the original massage from the sound produced in Morse-code, i.e. as tik (•) and Tok (-). Evan today, a Morse-code trained operator can easily converse 20-30 words per minute directly into English, from the sound produced, by the telegraph key. Since its invention, Morse-code remained a viable means of communication, even under difficult conditions. Besides sound,

Morse-code can also be transmitted using light, in the form of short and long light flashes, which are used to denote • – – –.

In case of emergencies, this practice is sometimes followed between ships on high sea. When no other form of communication is available ships transmit distress signals using light flashes. The standard international distress signal is SOS (••• – – – •••).

Since December 2003, Morse-code has included the symbol, @, which is represented as a combination of 'a' and 'c', i.e. • – – • – •, and this has been the only change in Morse code system, since the World War-II.

Initially, the telegraphic-messages were sent from the telegraph-office by specially trained operators using 'Morse-code'. The massages were known as 'telegrams' or 'cablegrams'. These terms have often been shortened as 'a cable' or 'a wire', message. Much later, telegrams were sent by the 'telex' network, which was a wired network of tele-printers. This was somewhat similar to the telephone network, but since it involved written massages, it came to be known as telex-messages.

10.5 Telegraph Machine

The process of telegraphy was done using machines, having electrical link between the transmitting and receiving centers for the messages. This was the conventional telegraphy. The word 'telegraph', in a restricted senesce, generally refers to an electrically transmitted telegraph massages. Wireless-telegraphy is more precisely known as the, 'radio-wave modulated massage transfer by on-off keying', as opposed to the earlier radio-technique using a spark-gap.

Morse, Along with two American physicists named, Joseph Henry and Alfred Vail, developed an electrical telegraphing machine in the year 1836, which was called the 'American Artist System'. This system sends pulses of electric current along wires which control an electromagnet, located at the receiving end of the telegraph system. A typical

Morse telegraphic key shown in **Fig.-2** was used for such telegraph system and it was called the 'U.S. model, J-38'. Very large number of such keys was manufactured in USA, during the World-War-II, and some of these remain in widespread use even today. With a telegraph key, the signal is 'on' when the knob is pressed, and 'off' when it is released. Length and timing of the dots and dashes are entirely controlled by the operator.

Ultimately the telegraphic key-machines were replaced by 'teleprinter'. These machines looked similar to a typewriter and the operation of alphabet keys produced Morse-coded signals, ready for transmission to the receiving station. The massage was finally reproduced in original alphabets and written or typed on paper, ready to be delivered.

10.6 Telegraphy spreads in USA, Europe and Aviation Industry

In the year 1837, William Cooke and Charles Wheatstone started the use of telegraphy in England, which used electromagnets in its receivers. In contrast to any other system of making sounds of clicks, their system used pointer-needles that rotated above a chart of alphabets to indicate the letters that were being sent. Later, in the year 1840, these workers built a telegraphic machine that printed alphabets from a wheel of typefaces struck by a hammer. This machine was based on their earlier telegraph machine and it worked well. Since they failed to create market for this system, only two such machines were ever built.

Later, in the year 1920, the use of Morse-cod was adopted by the aviation industry, in the form of 'aviation radio-telegraphic system'. During the initial period, the transmitting machines were bulky and the transmission system was slow and difficult to use. Yet, during the World War-I, the US-navy used Morse-code from airplane to ships.

Charles Lindberg made his first, trans-Atlantic flight, from New York to Paris in the year 1927. At that time the inter-continental radio-telegraphy was non-existence. Hence, at that time the historic event of landing of Lindberg's airplane in Paris, could only be conveyed to the press in USA, by then existing, slow sea-mail. Soon afterwards in the year 1930, the first trans-Pacific air-flight, from California to Australia, was communicated by its own radio operator, *via* radio-telegraphy to USA.

Before long distance and intercontinental telephone services became readily available and affordable for a common man, telegram services were very popular. Telegrams were often used to confirm business dealings, and unlike e-mail, telegrams were commonly used to create binding of legal documents for business dealings. The use of telegraphy has slowly declined, ever since 1970.

In India, the fees charged for sending a telegraph (irrespective of distance), is determined by the count of words. Hence persons, sending telegraphs, have a tendency to skip unnecessary details and what is communicated in a telegram has come to be known as 'telegraphic language'.

Indian telegraph offices have 'code numbers' for some commonly used massages of greetings for auspicious occasions and for condolence and sending it by a code number becomes cheaper. Compared to any private communication using a telephone, a communication by telegraph is considered to be more authentic, because it comes *via* the service of a government department.

The contents of a telegraph become an office record. Even there is a provision for sending 'telegraphic money-order', in case of an emergency.

Efficiency of telegraphic communication in India, during pre-second world-war period was demonstrated, when the King, George-V, died in England in the year 1939. This information was conveyed to the capital cities of all the countries, (including India) of the then British-Empire by telegraph. In a very short time the information was further relayed to different state capitals, cities and towns throughout the empire by telegraph. Thus the news reached on the same day.

In India telegraphy related public services have now been completely discontinued with effect from June, 2013. This became necessary because these were not frequently being used now and have become uneconomical.

10.7 Decline of Telegraphy and Arrival of Internet and Fax-Services

Starting with the twenty-first century and onward, there has been a faster decline of conventional telegraphy and now, the massages are more generally sent by 'Internet' in the form of 'e-mail', and by 'Fax-Service'. Such 'electronic-mail' was used by the US Defense Services even before the inception of world-wide internet, and it acted as a crucial tool in developing telegraphy.

Defense related research projects in USA, have played major role in developments related to internet. In the year 1965 a method of written communication, developed by the 'Defense Advanced Research Project Agency' (DARPA of USA), which was based on central switching system, was developed. Later on, this emerged as, what we know today as 'Internet' and 'Fax-service'. Of these two telecommunication methods, the internet service can be made available to any person, having a telephone connection and a 'personal computer'. For 'fax-service', one has to have a separate telephone connection exclusively for a 'Fax-Machine' **(Fig.-3)**. A fax-machine is generally left 'on' and keeps on recording, on paper, all incoming communications, without any personal assistance. Hence it has become a common facility in government and private offices.

The term 'fax' is a short form of facsimile and it means sending a document on telephone line. Fax machines, in various forms have existed, since the early 19th century, but the modern form of fax machine, was made in the mid-1970s. Digital fax machines became popular in Japan, where the competing technologies like the teleprinter already existed. It proved to be much faster, to hand-write 'kanji' (Japanese-script) type characters and to send it by fax. By mid-1980's faxing became affordable, and commercial fax machines became popular around the world.

10.8 Global Internet and Fax-Services

'Advanced Research Project Agency Net-work (ARPANET)', has been the other US-Agency, besides DARPA, which contributed towards the development of the modern, global internet, and e-mail services. International standards (RFC 561) for encoding e-mail messages were proposed as early as the year 1973. Conversion from ARPANET-technology to the Internet, in the early 1980s produced the core of the current services, in public domain. An e-mail sent in the early 1970s looked quite similar to a basic text message

Massachusetts Institute of technology (MIT) was the first to demonstrate, what later on came to be known as the 'Compatible Time-Sharing System (CTSS)' for internet, in the year 1961. CTSS allowed multiple users to log into the IBM-7094 computer, from any remote dial-up terminal. It also permitted storage of files online on disk. This encouraged users to share information in several new ways. E-mail service started in the year 1965, as a way for multiple users of a time-sharing mainframe computer to communicate.

10.9 Internet Services, E-Mail

Internet has been a radical break through, for communicating a written massage as well as any graphics and pictures. Internet operates by a digital transmission system, for which the routing is now completely decentralized. Large e-mail messages can be broken into smaller packets, which are then reassembled at the destination. As the Internet grew, it progressively used the faster digital carrier links, between different places.

E-mail is based on, simultaneously exchanging digital messages from a sender to one, or several recipients. Earlier condition of e-mail system, that both the sender and the recipient be online at the same time, is no more necessary. E-mail systems are now based on a 'store-and-forward' model. E-mail servers, at the sender's end shall accept a mail, in the form of a document. It shall then be forwarded, and shall be delivered from the stored messages, as and when the recipient desires and opens his mail. This means that neither of the users nor their computers is required to be online simultaneously. They need to be connected only briefly, to an e-mail server, for as long as it takes to send or to receive messages.

10.10 Conclusion

For verbal communication, land-line telephones and mobile phones are now widely used, throughout the world. Yet, at times the communication of a written massage becomes necessary. Telegraphy has been the oldest way of communicating written massages. Alternate communication technologies e.g. e-mail and fax-services have now emerged, which can send and receive written massages, besides photographs and figures, anywhere in the world. It has already been mentioned that like many other technologies, the conventional telegraphy is now 'out' in many countries including India. Millions of people, who do not own a PC at home, can still communicate their urgent massage *via* private fax and e-mail services available in most cities.

International Morse Code

1. A dash is equal to three dots.
2. The space between parts of the same letter is equal to one dot.
3. The space between two letters is equal to three dots.
4. The space between two words is equal to seven dots.

Fig.-1. Morse codes for alphabets and numbers.
Author, RT. Snoddgrass & VF Camp, Article on telegraphy, http:// en.wikipedia.org/

Fig.-2, A typical telegraph key

Author, RT. Snoddgrass & VF Camp, Article on telegraphy, http://en.wikipedia.org/

Fig. 3 A modern Fax Machine

Author, Jonnyl, Article on telegraphy, http://en.wikipedia.org/

Chapter-11

Decline of Fountain Pen and Emergence of Ball Pen

11.1 History: Invention of Language and Script

During his long history of existence on this earth planet, man has acquired many natural, evolutionary skills. These have helped him to become the master of this planet. One such skill has been the development of his vocal organs, by which he could communicate, in a specific language, with his fellow-beings. Man also developed a large brain to store more memory and information, than any other animal. Yet the human brain is not large enough to store all the knowledge, which accumulated over centuries, by the generations of his ancestors. For storing all such knowledge and information, which his ancestors, have gathered, man invented language and script and also developed the skill of writing, by which he could leave such accumulated knowledge, stored for his future generations.

Even a prehistoric, caveman, used a script of 'sign language' (pictures) on rocks of the cave, he lived in. Examples of these have been found widely scattered over the world, including in India.

Soon afterwards, man started recording his knowledge and information on portable sheets, such as wooden planks, bark of certain trees, parchment, woven cloth and handmade sheets of cellulose, which we now call paper. The discovery of 'Dead Sea Scrolls' in Israel is one such example.

Realizing the limitation of a sign language, for storing long accumulated information and knowledge, man devised scripts, which his other fellow beings could also read out. Development of such script may have started earlier than five thousand years ago, which is supposed to be the period, when sign language was being developed and used in Egypt. This has also been claimed to be the period, when the 'Vedas' were scripted in 'Sanskrit' language, which was based on alphabets, somewhat similar to those of today's 'Devnagri Script' of Hindi. Most of the current languages of Eurasian continent have a common origin (Sanskrit), though the scripts in which they are now written may be quite different. The script used for writing various regional languages of India today, also differs considerably.

11.2 Ink and Pen: The Media for Writing

For recording information, there was a need of some type of liquid-ink. Primitive man may have tried to use a variety of available plant pigments, for painting and writing, but soon he realized that the color of most of the plant pigments faded away and did not produce a lasting effect. Therefore he considered using mineral colors, which do not fade away.

Man was familiar with black soot, which he has seen, being produced during incomplete combustion of oils and fats, which he burnt to produce limited amount of light, during the nights. He also observed that the black color of soot did not fade away with time. This became a handy material for making black ink as well as a decorative pigment for his body. He could also make inks of certain other colors, from minerals, e.g. red from ocher (iron-oxide) and green from chromate-ore. Using nano-particle sized powder of these colored materials, he could make storable, aqueous dispersions, stabilized by added gum-acacia (stabilizer for hydrocolloids), which we now call, 'Black Indian Ink'. Even today 'black-ink' is prepared from soot (carbon nano-particles), which we know as 'carbon black'.

During ancient days, the inks thus prepared were stored in large brass inkpots (*Dawat* and *Kalamdan,* in Hindi); with a band of cotton fiber were placed inside the inkpot. This prevented any damage to the sharp writing tip of a wooden pen (*Kalam*), by striking against the hard metallic surfaces of the inkpot.

Such brass inkpots are available, even now, with many old and educated families in India and these have now become antiques. '*Kayasthas*' are one of the important Hindu communities, who were long associated with bookkeeping services in kingdoms and '*Jagirs*', of the Hindi speaking regions of India. Being associated to such services, where lots of writing work was involved, they had a practice of worshiping their writing material (dip-pen and ink-pot), which provided them living. Such worshiping was done, every year on the day of '*Bhyia-Dooj*', after *Diwali* festival and it was called '*Dawat-Pooja*' festival.

11.3 Traditional Inks for Penholder

Unlike today's inks, made from synthetic organic dyes, inks made from mineral pigments and soot did not fade away even on long exposer to air and light, for centuries. Thus we have many hand-written old books, Holy Scriptures and other documents, preserved in various museums in India and abroad, which have not faded over long periods of time. In writing of the Holy Scriptures, in India, the usual practice was, not to bind the pages in the form of a book. Written sheets of paper were arranged serially, and these were generally wrapped in a cloth.

11.4 Natural and Fabricated, Dip and Write Pen

Before any other type of writing pen was developed or designed in Europe, the sharp-ended quill was used as a natural 'dip and write' pen (**Fig.-1**). It is probable that the alphabets thus scripted were large, but then, such scriptures were only meant to be the part of collections at home, libraries and museums. In India, where bamboo-type of plants grew, very thin, hollow bamboo type sticks (*Kalam*), of about 5.0mm diameter was sharpened, at one end, into a 'nib-like' shape, using a penknife. These were referred as *'Lekhini'* (Hindi) or *'Kalam'* (Urdu and Persian) and these were used as 'dip and write' pen.

When East India Company was established, specially carved wooden 'pen-holder', to which a changeable metallic nib could be attached at one end, was introduced in India. *'Kalam'*, or a penholder with a nib was the most popular instrument for writing, till almost the time of independence in India.

We can now guess that the famous, Hindi fiction writer like Munshi Prem Chand and many others, must have spent a large part of their life, using dip and write type pen-holder with nib or a *'Kalam'* for making their enormous literary contributions.

In the early twentieth century, some imported fountain pens were available, but only to a highly privileged class of persons in India. Thus high government officials were provided with 'Parkar-51' fountain pens. Most of the educated population at that time was contended with dip and write type arrangements.

11.5 Ink-Pot and Pen in Schools and Offices

In offices and educational institutes, inkpots were provided on desks, with a regular refilling of ink, while dip and write type of nib-pens were generally assigned to individuals. This was the time when the offices were judged from the type of office stationary, they had, and one of the most prominent stationary was a *'Kalam-Dan'*, with decorative glass inkpots (generally blue and red inks) and a variety of pens and pencils.

In schools, there used to be a practice of using different nibs for writing in different scripts, i.e. Hindi (*Nagari*), Urdu (*Persian*) and English (*Latin*). For Urdu writing, nibs used were such that it could change the thickness of a line in an alphabet from one end to the other, and create a dot, which was not circular, but square.

11.6 Limitation of Dip and Write Pen

One major handicap of using dip and write system was that it was slow and an inkpot was not always available to a person, when he was away from his office or home and while traveling. Inventors were then busy to design a pen, which will have a built-in storage for ink, and which can be used anywhere. Need for such an invention, resulted in the form of fountain pen.

11.7 Fountain Pen, What is it?

A fountain pen is also a 'nib-pen' **(Fig-2)** but unlike a 'dip and write pen', it is provided with a built-in storage (reservoir) for an aqueous-solvent based, liquid ink. While writing on a paper, ink is continuously drawn from the ink-storage, and fed to the nib *via* a feeding arrangement **(Fig.-3)**, which is based on capillary action and gravity. It works as if the nib is continuously kept dipped in ink. The built-in ink-storage of a pen has to be periodically refilled, manually. This ink filling arrangement used to be different in different models of fountain pens, and was based on the choice of the manufacturers. In some cases ink was filled, using an eye-dropper or a syringe into a reservoir, which was a cylindrical plastic tube, closed at one end. The other end of the tube was screwed to a water-tight, plastic nib-holder. In other cases the ink was sucked-in, *via* nib-capillaries into a flexible rubber-tube reservoir by creating vacuum. Pre-ink filled cartridges were also used in some other fountain-pens.

11.8 Historical Development of Fountain Pen

According to a historical record, a Caliph of Maghreb, in the tenth century A.D., ordered for the development of a pen with a built-in ink reservoir. But its construction details are not known now. Till middle of nineteenth century the development of a dependable fountain-pen was very slow. This was partly due to incomplete understanding of the role of air pressure inside the ink reservoir, capillary action and gravity, on the flow of ink, onto the nib in a fountain pen.

On May 25, 1827, a Romanian inventor, named Pettache Poenaru, got a French patent for the invention of the first-ever fountain pen, with a replaceable ink cartridge. From 1850 onward there was a steadily increasing stream of patenting designs, for fountain pens and their commercial production.

Fountain pen became a widely popular writing instrument, only when three key inventions, related to it were made. These involved-

(1) Use of iridium-tipped gold nib,
(2) A hard rubber, or plastic body, and
(3) The availability of specialty, free-flowing and quick drying, fountain pen inks.

By 1850s these conditions were being met by most of the fountain pens manufacturers.

11.9 Specialty Nib for Fountain Pen

Nib used in fountain pens, has to be specially designed. Just as a dip and write pen, the fountain pen nib also narrows down to a point where the ink is transferred from the nib to the paper. Like a dip and write nib, fountain pen nib also had a vertical slit cut down in its centre, to convey the ink down to the tip of nib, by capillary action. There was also a need for having a circular hole in the slit, to replace the flowed-out ink, by air in the pen's ink-reservoir. This hole also acted in relieving any stress, which was developed due to pressure exerted on the nib-point, during writing. Thus, it prevented the nib, from onward cracking up longitudinally from the end of the slit-hole. Such longitudinal cracking could arise due to repeated flexing of the nib during writing.

The design of a modern fountain pen nib may be traced back to the original gold nib, which had a tiny fragment of ruby attached to it, form its wear-point. When more of the platinum-group of metals (ruthenium, palladium, osmium and iridium) were discovered, a small quantity of iridium was alloyed to make iridium-tipped, gold dip-pen nibs during 1830s. Today, the fountain pen nibs are usually made of stainless steel or 14 karat gold.

Gold is considered to be the optimum metal for its flexibility as well as its resistance to corrosion. However the corrosion resistance of gold is not an issue now, because of the availability of much cheaper, stainless steel alloys, and due to the availability of far less corrosive inks. Gold nibs are tipped with a hard, wear-resistant alloy that typically uses metals from the platinum group. Though the tipping material is often called iridium, but there are very few, pen-makers, who use alloys containing iridium for fountain pen nib tips. Steel nibs also have harder nib tips, because un-tipped steel points will wear more rapidly due to abrasion. Although the most common nibs, ended in a round point of various sizes (fine, medium and broad), several other nib shapes are also available.

11.10 Innovative Fountain Pens

The period starting from the year 1880 and onward is considered to be an era, when the mass-production of high quality fountain pen started. Further the period, which followed the World War-1, saw introduction of some of the most notable models of fountain pen. Some of these were Parker, Duofold, Vacumatic, Sheaffer and the Peliken-100.

Some of the dominant American manufacturer of this pioneer era was Waterman Company of New york and Writ Company of Bloomsburg, in Pennsylvania. Of these the Waterman Company became the world fountain pen market leader, during the early 1920s.

Most of the initial development problems related to fountain pen have been solved by now. Now the attention was directed towards designing of a convenient, self-filling fountain pen and to overcome the problem of ink-leakage during writing. Self filling fountains began to arrive in the market by the end of the nineteenth century. Most successful of these was Conklin's 'crescent-filler', followed by Waterman's 'twist-filler' model.

Walter, which was Sheaffer's 'lever-filler' fountain-pen, was introduced in the year 1912, and it got a runaway success, compared to many contemporary makes. This, Sheaffer's pen competed well with Parker's 'button-filler' fountain pen. Montblanc and Visconti were among the other fountain pens, of that period, which became more of a status symbol, rather than a part of regularly used writing media. Most of these lever-filling pen models used interchangeable steel nibs, and being cheaper, they became popular among masses.

11.11 Fountain Pen Industry in India

Many of us might be under an impression that manufacturing of fountain pen and ball-pen, in India has been a post Second World-War activity of mid 1950's. In an informative and very recent article by Nitin Pai, in an airport magazine 'Mint Lounge' (August, 20, 2011), a very interesting history of handmade fountain pen industry in the Rajahmundry town of Andhra-Pradesh (India) appeared. According to this information, during mid-1930's, and in response to the 'Swadeshi' call from Mahatma Gandhi, one Mr. K.V. Raman, started an industry to manufacture 'hand-made' fountain-pens in Rajamundry town. This tradition has continued in some towns of Andhra-Pradesh even today, where manufacturing of handmade ball-pens and fountain pens is still carried. These, 'Swadeshi' pens were evaluated by several pre-independence leaders and freedom fighters, including Mahatma Gandhi and Jawaharlal Nehru, who found these to be 'as good as the imported pens of leading international manufacturers' of that time.

A favorable factor for this industry in AP, was the availability of a natural plastic (polymer) ebonite. Ebonite is a polymer [poly-isoperenoid, $(C_5H_8)_n$], structurally similar to natural rubber, but due to its being a *trnas*-isomer (rubber is a *cis*-isomer), it lacks the elasticity of rubber. Ebonite could easily be molded into fountain pen tubes.

This industry in Andhra Pradesh has now extended to some other towns also and fakes pens of brand-name e.g., Montblanc's and Waterman's are now made here. This has come as a strong challenge to similar industries in China, which makes such flashy China-made fake fountain-pens.

11.12 Decline of Fountain Pen and Emergence of Ballpoint Pens

During the period 1940-1950, fountain pens retained their dominance as instruments of writing. During mid-nineties ballpoint pens appeared on the scene. Initial models of ballpoint pens were expensive. They leaked, and in some of them the flow of ink was also irregular.

After the Second World-War, new manufacturing capabilities in the USA provided reliable and economical technology for manufacturing of ballpoint pens. This technology, which resulted from a series of new trial, was based on modern physics and chemistry. For this, high precision and a series of experimentation became necessary. In its early stages of development, ballpoint pen had several failures in design. This resulted in not getting worldwide patents for ball pen. Ultimately such ballpoint pens were developed, which were cheaper, and commercially viable products.

11.13 Earlier Attempts for Designing Better Ball-pen

The first patent, on ballpoint pen was issued on 30th October 1888, to John Loud, who was a leather tanner by profession. Loud's basic interest was to make a pen, suitable for writing on his leather products. This could not be done using, regular fountain pens. A ball pen was designed by Loud, which had a tiny, rotating steel ball, which was held in place by a socket at the writing end of the pen. The name 'Ball-Pen' was coined for this new pen. This pen could be used to mark rough leather surfaces, but it proved to be too coarse, for writing on paper, hence it was not commercially exploited.

During the first-half of 20th century, several improvements were made in, then available ball pens. In the year 1907, Slavoljub Eduard Penkala invented a solid-ink fountain pen, while a German inventor, Baum got a patent on a ballpoint type pen in 1910. Van Vechten Riesburg got a patent on a ballpoint pen in the year 1916, which was based on the use of a thin ink-filled tube, with one end blocked by a tiny ball in a socket. This ball could not slip into the tube or fall out of the pen. As the pen was drawn across a paper, the ball spun, and ink clung to the ball continuously. This resulted in normal writing on a paper, and the origin of its name as ball pen. Still none of these earlier designed ballpoint pens delivered ink uniformly on the paper.

This was due to the fact that either the ball socket was too tight, for the ink to reach the paper, or it was too loose, and permitted large amount of ink to flow past the ballpoint tip, which resulted in writing paper getting smeared with ink.

A Hungarian newspaper editor, Laszlo Bíró, being frustrated by the time wasted in frequent refilling up of his fountain pen and cleaning up of smudged pages, started to work on a modified fountain pen. The sharp tip of his fountain pen often tore the writing paper. Being associated to newspaper printing industry, Bíró had noticed that inks used in newspaper printing were quick drying and hence these inks left the printed paper dry and smudge free. Based on these observations he considered to make a pen-ink similar to that used in printing of news paper.

Since, the viscous printing-ink would not flow into a regular fountain pen nib, Bíró, with the help of his brother George, who was a chemist, began to work on designing new types of inks and pens. Bíró fitted his pen with a tiny ball in its tip that was free to rotate in a socket. The ball rotated as the pen moved on the paper, picking up more ink from the ink cartridge and leaving it on the paper. Bíró filed a British patent on 15 June 1938, for the pen he has designed.

11.14 Fountain-pen Manufacturers turn as Ball-pen manufacturers

It was mentioned earlier, that the use of fountain pen started declining after 1950 and ball-pen took its place from then onward. In India, initially the ball pens had to be imported, but soon their production was started. Some very simple, 'use n through' type ball pens are now mass-produced and these are used in schools and colleges. Since these ball-pens used a refill, they could be used until they are misplaced or lost. More sophisticated ball-pens are now being made for persons, who are concerned with regular writing.

Considering the decline of fountain pen market and steadily favorable inclination of writers towards ball-pen, the manufacturers of fountain pen soon realized that the future lies in manufacturing of ball pen. Most of the international manufacturers are now manufacturing ball pen **(Fig,-4)** with an earlier name of their fountain pen. These manufacturers have also come up with presentation set of a ball-pen and a fountain pen in a box. However it has been observed that the person, who receives such gift-presentation, generally starts using the ball-pen, while the fountain pen keeps on lying in the box and perhaps may not be refilled with ink even once.

Thus there has been a real decline of fountain pen and it is being replaced by ball-pen.

Fig.-1 Inkpot and Quill (used as pen)
Wicipedia article on Inkpot and quilt,
http://en.wikipedia.org/

Fig.-2. Parker Duofold Fountain-Pen, 1924 and 1928
Photo, Wikimedia Commons, http://en.wikipedia.org/

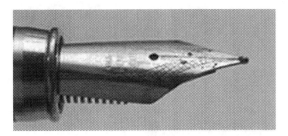

Fig.—3, Fountain pen nib labeled (Iridium tipped)
Photo, Wikimedia Commons, http://en.wikipedia.org/

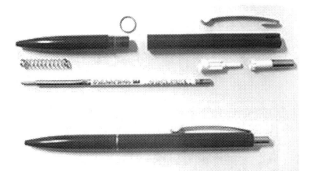

Fig.—4, A typical Ball-pen, and its components
Photo, Pavel Cork, Article on Ball-Pen, http://en.wikipedia.org/

Chapter-12

❖

Decline of Steam-Locomotives and Emergence of Diesel and Electric Locomotives

12.1 Defining a Locomotive

The word 'locomotive' is derived from two Latin words. In Latin language the term '*loco*', (it is an abbreviations of '*locus*'), which stands for, 'from a place', while '*motivus*', means 'causing a motion'. When the mobile steam engines were first introduced, in the nineteenth century, the word 'locomotive steam-engine' was coined for them. A locomotive engine is a vehicle, which is used by the rail-transport organizations, to run trains, and it provides the moving power for a train.

It may be well understood that a locomotive engine, by itself is not provided with any space or capacity, to carry a payload of either passengers or goods. When a locomotive is connected to a train, it only serves to move it, along with its payload of coaches, which carry passenger and goods on the rail-tracks. Hence the sole function of an engine is to move a train along its tracks.

However, there are some self-propelled, payload-carrying trains also. These do not need any locomotive engine. These are frequently referred as multiple-unit, motor-coaches or rail-cars. The use of these self-propelled vehicles has recently very much increased. These inter-metro, passenger trains are generally referred as 'metro-trains or simply 'metro'. These trains are not basically meant to carry freight between different places, in a city, but only inter-city passengers, particularly of large city. Power-cars are vehicles, which provide motive power to haul an unpowered train, but these are not generally considered locomotives because they have payload space, and generally not detached from their trains.

Traditionally, a locomotive pulls a train from the front only. Due to an increased overall load, some trains also have engines, one each for push and pull operation. The front locomotive pulls the train in one direction, while another engine at the back pushes it, in the direction of pull. In India such arrangements are also seen on certain mountain tracks, where the power of one engine is insufficient to pull a train on a steep slope.

12.2 Inter-Conversion of Different forms of Energy

Energy has been defined as the power to do work. Heat is one of the many different forms of energies. In case of solids, the heat energy is a measure of the rate of vibration of the particles, (atoms or molecules) constituting a solid, about their mean position. Such vibratory motion, completely ceases at the 'absolute zero degree temperature' (-273°C). Particles of liquids and gases, which are collectively known as fluids, additionally have energy due to, 'rotational, irrational and translational' motions. These motions are increased due to heat, resulting in an increased temperature.

Translational motion of the particles in a gas is random, i.e. having variable velocities and not confined in a particular direction only. Hence, when a well-defined and unidirectional motion is desired for doing certain mechanical work, the whole of the translational motion (energy) of hot-gas particles is not available for conversion into mechanical work.

This is the 'Second Law' of thermodynamics, which states-

'Heat (energy) cannot be converted into work (mechanical energy), without a loss (compensation)'.

The fraction (or the percentage) of the total heat energy which is converted into work is called the 'efficiency' of a machine (engine). Theoretically if whole of the heat energy were to be converted into mechanical energy (work), the efficiency shall be one, or 100%.

Unlike the conversion of heat into work, the reverse conversion is possible at nearly 100% efficiency. This is also true of conversion of electrical energy into heat or mechanical energy.

12.3 History of Locomotive Seam-Engine

Initial development of locomotive steam engines, took place in the United-Kingdom.

During the early nineteenth century, the only industrial fuel known to man was coal. The application of petroleum based products, to be used as a fuel was made much later. Internal combustion engines made now were not possible without liquid or gaseous fuels. With coal as the sole fuel available, only steam engines were feasible and an era of stationary steam engines had commenced even before loco steam engines.

The purpose of an engine is to convert the chemical energy, stored in a fuel, into mechanical energy. This takes place *via* heat energy produced on combustion of the fuel. In such conversion of stored chemical energy of a fuel, (*via* heat) into mechanical work, there is always some loss of energy.

In the year 1804, the first locomotive steam engine, which was built by Richard Trevithick, was used to haul a train, along iron rails, then being used by a tramway of the Penydarren Iron-works, at Merthyr-Tydfil in Wales (United Kingdom). At a time, this locomotive, could haul, about ten tons of iron-ore and seventy passengers on a five wagons train. The distance it covered, was fourteen kilometers. Even this load was considered too heavy for the cast-iron rail track built for a tram-way at that time, and hence it was discontinued. Thereafter Trevithick built more locomotives, one of which ran at a colliery in Tyneside in northern England.

The first commercially successful steam locomotive built was Matthew Murray's, rack locomotive, named 'Salamanca'. It was built for the narrow gauge track, for the Middleton Railway in the year 1812. An year later this was followed by the 'Puffing Billy' built by Christopher Blackett and William Hedley for the Wylam Colliery Railway, which was the first successfully running locomotive. Puffing Billy is now on display in the 'Science Museum' in London, and it is the oldest locomotive in existence.

12.4 George Stephenson's Rocket

In the year 1814 George Stephenson **(Fig.-1)**, inspired by the early locomotives of Trevithick and Hedley, persuaded the manager of the Killingworth colliery, where he was working, to allow him to build a steam-powered, locomotive engine. He built the 'Blücher', which was one of the first successful flanged-wheel adhesion locomotives.

Since then, Stephenson played a pivotal role in the development and spreading the use of steam locomotive engines in U.K. His designs were later on further improved by the work of other pioneers in this field. In the year 1825, Stephenson built a locomotion engine, for the Stockton and Darlington Railway, in North-East England. This engine became the first steam locomotive of a public railway company. In the year 1829 he built another engine, named 'The Rocket', **(Fig.-2)** which competed in, the Rainhill Trials for engines and won it.

The first inter-city passenger train, owned by the 'Liverpool and Manchester Railway', opened in the year 1830, making an exclusive use of steam power for both passenger trains and freight trains. This success led Stephenson to establish his own company, and the most prominent builder of steam locomotives, which were used, at that period by railways in the United Kingdom. Earlier to this, in the United States of America and most of the European countries, horses were used to haul some passenger carrying trains. To reduce the friction due to roads, such horse driven trains used to run on specially built steel track or the 'rails'. This practice of using steel tracks or the rails was also followed for steam locos, from which the term 'railways' has originated.

12.5 Limitations of Steam Engines

A steam engine uses coal as the fuel, which is burnt in a specially designed furnace and the hot gases, or the combustion products are used to produce steam in copper-tube water-boiler, at a very high pressure. Steam, thus produced is introduced into a piston-cylinder system of an engine and linear motion of the piston is made to rotate the wheels of the engine *via* a connecting-road system.

In **Fig.-3** is shown a much later developed and high powered modern steam engine. A steam locomotive generally requires a preparation of 2-4 hour in a loco-shade, where they are kept for maintaince and servicing, before commencing on a journey. It has to take enough coal to last for a journey to another station, where it can be refueled with further supply of coal and water. Coal as a fuel yields far less energy (calorie) for a given mass and produces large amount of smoke and fly-ash. In a boiler based, piston-cylinder type machine the energy efficiency is very low (~15%).

Another important requirement of a steam engine is large amount of good quality, soft water for producing steam in its boiler. Poor quality of hard water can result in scale formation. Hence there have to be many watering stations at regularly distanced stations, on its route.

It has been mentioned, in the history of British-Railways that a train, named, 'Flying Scot', in the UK could replenish its supply of water, from a water-channel, laid along its track, but still it had to stop for a fresh supply of coal. These factors limited the distance, which a train with a steam engine could cover at a time, without stopping. In spite of these limitations, steam locomotive engines did a great job for almost a century, for serving passengers in their land journey and hauling goods, throughout the world..

In many countries, including India, there are 'Rail-Museums' where old rail-engines and rail-coaches are preserved. On certain special occasions, an exhibition-train, constituted from these antique coaches and engines is also run.

12.6 Indian Railways

India celebrated the 'Centenary of Indian Railways' in the year 1953, which means that the first railway in India was established in the year 1853. Thus India was not much behind Europe or USA in establishing rail-transport. This was understandable from the fact, that due to its huge natural resources, Imperial-India had a large export potential. Many private rail companies came forward to establish a huge, interconnected and well spread a net-work of railways in India. Initially there were three different rail-gauges, i.e. the narrow-gauge, broad-gauge and a special, narrow gauge, which operated in hilly regions only.

This necessitated establishment of loco-workshops, by the rail companies, for the maintaince of steam-engines, which were imported into India at that time. Much later and after independence, 'Chitranjan Locomotive Company' was established in the 'Public Sector' in India. This company later on expended its activity to start the manufacturing of diesel and electric rail engines also.

12. 7 Bidding Fare-Well to Steam Engines

In North America and in many European countries, many heritage railways still run steam-engine powered trains (*Chhuk-chhuk train*) during the tourist season. These are largely aimed at attracting tourists, and travelers who use railroad travel as a hobby, and who call themselves 'rail-fans'. Some narrow gauge, steam powered trains in Germany, have become a part of the regular public transport system. These trains have a year-round running schedule, using steam engines for their motive power.

Steam locomotives were in commercial use in some parts of Mexico till late 1970s. In China, steam locomotives were being used regularly until the year 2004. This has been due to the fact that coal is a much more abundant as fuel resource in China. Crude petroleum, which is the source for diesel oil and other automobile fuels, is largely imported into China.

In India, major transition from steam-locomotive powered trains, to diesel and electricity powered trains took place in the 1980s. Currently some heritage trains in India, particularly in the high altitude, mountainous region (Ootty, and Kalka-Shimla, narrow gauge rail-links) still have regular rail service, where steam engines are used, though at the high altitude, and due to a reduced air pressure, steam engines are less affective, compared to diesel engines. In South-Africa, steam-locomotives were routinely used in passenger trains, till late 1990s. Here again some of these are now reserved for tourist trains. In Zimbabwe steam locomotives are still used on shunting duties, around Bulawayo and on some regular freight services.

12.8 Diesel and Electric Locomotives

During the mid-20th century, when electric and diesel-electric locomotive engines **(Fig.-4)** were designed, they gradually started replacing steam engines.

Diesel locomotives have internal-combustion engines similar to those used in large trucks, buses and ships, but these have a much higher horse-power. These engines do not drive the loco directly. It is used to power a generator, which supplies electricity to electric motors, which are mounted between the wheels of an engine.

Steam locomotives were far less energy efficient, compared to their more modern diesel counterparts. Being internal combustion engines, these have efficiency of 30-50% of thermal energy of the fuel used.

Steam engines require much greater manpower to operate and to service them. It has been estimated that the cost of fuel and the salary of the staff of railways in UK, when running on coal, was nearly two and a half times that of now diesel powered machines. Also the daily mileage achievable by the steam locomotives was far lower, when compared to diesel engines.

Non-steam technologies have become even more cost-effective, due to the increase of labor costs. This particularly happened during the post 'Second World-War' period. During the period 1960-70, in most of the western countries, the steam locomotives were completely phased out and replaced by diesel engines. Extensive introduction of electrically operated passenger train services have also taken place. Freight locomotives, however took some more time to be replaced, but in most of the western countries the steam locomotives have now become a thing of past.

12.9 Diesel Loco-Engines

It has been mentioned that unlike the steam engines, diesel-engines are internal combustion engines, based on liquid fuel. These have multiple cylinder-piston engines. In an internal combustion engine, a mixture of atomized-spray of a liquid fuel, dispersed in air, is ignited to produce a mixture of hot gases, in the confined volume of a cylinder, which is closed at one end. Gaseous combustion products being confined in a closet space, pushes the piston to perform work. Diesel engine is used to generate electricity, which in turn powers electric motor.

Because of the increased availability, the use of liquid fuels has now very much increased. This has resulted in economy and more convenience in their use and their portability coupled with storage. Hence the diesel engines are now being used, not only in trains, trucks and ships, but also as stand-by for the generation of electricity. Unlike steam engines, diesel loco-engines do not need more time for initial starting and can store more fuel. As a result, these can cover longer distances without stopping for refueling. The higher limiting speed of diesel engines is far more than that of a steam engine.

12.10 Electric Loco-Engines

Electrically driven locomotives are much smaller, compared to the diesel engines, because these do not have large diesel engines. These are directly powered by electricity drawn from an over-head power line. Electric motors, mounted between the wheels, turn the wheels of a loco-engine **(Fig.-.4)**.

Electric loco-engine is a sort of misnomer. Engines are machines, used to convert heat energy from a fuel, into mechanical energy, which is then utilized for doing work. The so called, electric loco-engines do not use any fuel. They have electric-motors, running on electric energy (power), which is generated in a land based power generating station, and supplied to the rail engine by an over-head power-line for loco-motion. Such passenger trains are popular for city transport system or the metro-rail system. These are completely non-polluting.

12.11 Metro-Rail and Tram Services

With expending cities, it has become necessary to have metro-rails, which run within a city and its suburbs. At present, in India there is a tram-service only in Kolkata. Till mid-1950's there were tram services in Mumbai, Madras city and Delhi also. Later on these were discontinued. Tram services are generally slow and being self-propelled, trams do not have engines, but use over-head electric power for its running, *via* electric motor. Steel tracks for trams are laid on the sides of normal traffic-roads in a city and these, frequently obstruct smooth running of other automobiles and private transports in a city. This has been the major reason for removing tram services in many cities in India and other countries.

There has been a long history of metro-rail, particular in Mumbai and Madras. These suburban or the metro-rails do not need any locomotive engine, but these are referred as multiple units, motor coaches or railcars. To reduce traffic jam in cities, the use of these self-propelled vehicles has very much increased now. In many parts of the western-world, metro-train services have been built under-ground. In London it is referred as 'Tube-Railway'. In India, where new metro-rail services, (e.g. those in Delhi, Bangalore, Hyderabad, Jaipur and Mumbai), have been introduced, the rail tracks have been built, above ground on RCC-pillars. This has been done without reducing the existing ground-road space. Metro-trains are run from the over-had electric power supply and do not need locomotive engine in a general sense.

To sum-up, there has been a gradual and effective transition in transport services, from steam locomotive to diesel and electric metro-trains.

Fig.-1. George Stephenson, Inventor and Engineer
Wikimedia Commons, http://en.wikipedia.org/

Fig.-2 Drawing of a Locomotive constructed in 1816 by Stephenson for the Killingworth Colliery
Wikimedia Commons, http://en.wikipedia.org/

Fig.—3. A typical, modern steam locomotive engine
Wikimedia Commons, http://en.wikipedia.org/

Fig,—4. A typical diesel locomotive engine
Wikimedia Commons, http://en.wikipedia.org/

Chapter-13

Electrical Energy and Electronic Technologies

13.1 Introduction

In nature, electricity is generated during a thunder-storm, which can be seen as a flash of lightening in clouds. This can have a voltage of several millions of volts, while a single, manmade battery only produces 1.5-2.0 volts. Static electricity, is produced by friction of synthetic cloths, and makes these to stick to our body. Sometimes it produces crackles, when combing hairs, but the amount of electric-charge, thus produced is very small.

A large electrical power-house can have capacity to generate hundreds of million watts of electric energy. An ordinary incandescent (filament) bulb consumes between 25-200 watts of electricity to glow and produce light and heat. Electricity is needed for running a fan and for broadcasting of news as well as for receiving a TV or a radio program. Thus we note that electricity has now become a prime necessity in man's life.

Presently, the world-over generation of electrical power (energy) is mainly, based on combustion of fossil-fuels e.g. coal, natural gas and petroleum based products to produce thermal energy, and convert it into electricity, *via* steam turbines or engines.

Fossil-fuels, such as coal, petroleum and fuel gases, which are our current sources of thermal energy, were formed on the planet earth, over past billions of year, from anaerobic decaying of dead plants and animals. Due to certain natural calamities, these got buried and have been laying buried, deep inside the earth's surface and sea beds for millions of year. Fossil fuels are none, regene'rable sources of energy. Large scale utilization of these fuels on industrial scale only started, about 2-3 centuries back. A large scale, commercial exploitation of fossil-fuels started due to several, science based technological developments, during past several decades.

Some technocrats and scientists have opined that a long term and increasing use of fossil fuels can be detrimental to the earth's delicate eco-system. It can result in global warming and ultimate change in the present climate cycle. This can ultimately result in rise of sea level and the seas engulfing some of the present, land-habitat for man.

Looking to an ever-increasing demand for electricity and to produce more of it, there is a need to use natural and renewable sources of energy e.g. the wind, sea-waves and solar-energy.

13.2 Sources of Energy and Generation of Electricity

Science has always been the predecessor of technology, and this has also been true of the technology for electricity (power) generation and its use.

Electrical energy is more important than other forms of energy, because this is the only form of energy, known to man, which after generation at a particular place, can be transmitted to other places, which are thousands of kilometers away. It can then be used at those distant places, by reconverting into mechanical energy form. Hence, other forms of energies are invariably converted into electricity, before use.

Next to thermal power generation; currently the hydro-power generation is the second largest power source, being exploited all over the world. Unlike fossil fuels based electricity generation, there is no hazardous impact of generating electricity using hydropower on the environment. After generating electricity, water from a hydropower generation plant is invariably available for irrigation. Nuclear energy, in spite of certain hazards, is emerging as a big source of energy, which is a form of green energy. As yet, there is still a need to minimize hazards in generating nuclear energy.

Though hydropower is a green energy source, its main limitation is that it can only be confined to certain hilly regions, from where mighty rivers originate and provided these can be dammed. The flow of water in a river, after generation of electricity is used in agriculture. Just like hydropower generation, some more 'green energy sources' are being explored as alternative energy sources. These include wind-energy, solar-energy, sea-waves and many more. These energy sources, unlike the fuel based thermal-power are renewable and almost un-exhausting sources of energy. Due to the current technological constrains, the utilization of these green energy sources have been limited. Sun-light as an energy source is the largest and most wildly distributed and available form of energy, all over the earth.

Even during early twentieth century, T. A. Edison for-saw sun as an energy (electricity) source. He is reported to have commented to his financers, Henry Ford and Firestone in the year 1930-

"I'd put my money on the sun and solar energy. What a source of power? I hope we don't have to wait, until oil and coal run out before we tackle that."

Though not the animals, but plants do make an excellent use of solar energy to photosynthesize all the materials, from which they are built and grow. Ultimately the plant products (directly or indirectly), form food (energy source) for all animals, including the man.

Over millions of years, the plants have been synthesizing organic material, which due to natural calamities got buried inside the earth, where it has decayed under anaerobic conditions, to produce coal and petroleum (fossil fuels).

13.3 Direct Current Electricity (DC)

Commercial generation and distribution of electricity in the USA commenced during late nineteenth century, when Michel Faraday's invention of dynamo made it possible to generate electricity. Two of his contemporary scientists cum technocrats, namely Edition and Nicola, worked on commercialization of electricity. Tesla was yet another technocrat, who worked on commercial production of electricity. Tesla was in favor of setting up of an electricity distribution system, involving DC-electricity. In contrast, Edison was in favor of AC-electricity. Technical feasibility was in favor of generating AC-current, which ultimately became available for domestic and commercial uses.

One of the simplest sources of electrical energy is a rechargeable, storage battery **(Fig.-1)**. A battery consists of two electrodes, called the cathode and the anode, which are in contact with certain electrolytes. An electrolyte consists of certain chemicals, in the form of form of aqueous solution or pastes (gels). Due to redox-chemical reactions taking place at the electrodes a potential difference is created between them. When the two electrodes are joined externally, *via* connecting wires to an electric appliance, e.g. a small bulb or a heating coil (resistance), electrons start flowing through it. This results in converting the electrical energy into light and heat forms of energy. By a convention, the flow of current is presumed to be from the anode to the cathode, while the flow of electrons, actually takes place in opposite direction.

If the potential difference (Volt) between the two electrodes is 'V' and the bulb-filament resistance (Ohm) is 'R', then the current 'C' (Ampere) flowing shall be related as-

$$V/R = C, \text{ This is the Ohm's Law.}$$

Such generated electricity is called 'Direct Current' or DC, which means that the flow of electrons is unidirectional **(Fig.-2, Red line)**.

After certain amount of battery use, the chemical reactions taking place at the electrodes, nearly ceases and the battery needs to be recharged. This is possible by applying a slightly higher DC voltage, through the electrodes, in opposite direction, which results in reverse chemical reactions taking place at cathode and anode compartments and regeneration of electrolytes.

The amount of electricity, that can be generated from a battery is much smaller, (1.5-2.0 volts) compared to that needed as a public utility service and hence it has a limited use. In many electronic equipments, where DC-current is needed, AC-current available from the commercial supply line can be converted into pulsating DC **(Fig.-2, Red line)** using a rectifier and a transformer. Thus the need for a battery is eliminate.

Some technologists still think that commercial generation and distribution of DC-electricity may become a future choice. This, however, is not likely to happen, at least in near future, and may take many more decades or even centuries. The money and efforts needed for such a change is likely to be enormous.

The increasing role of DC has been attributed to the ever increasing applications of semiconductors in electronic technology, which is based on DC-current. Whenever an electronic appliance, converts AC into DC, there is always some loss of energy in the form of heat produced in the transformers and rectifiers, used for such conversion.

It has been mentioned earlier that among the group of pioneers, who developed the technology of generation and distribution of electricity, Edison was in favor of AC electricity, while Nicola and Tesla, favored DC form of electricity.

13.4 Alternating Current Electricity (AC)

Unlike the direct current (DC), where the flow of electric charge (electrons) is in one direction only, in case of alternating current (AC), the movement of electric charge periodically reverses its direction. This gives a waveform (sine-curve) representation to an AC electricity power circuit. AC form of electricity is generated on commercial scale, in power-houses using machines called 'dynamos' or the 'generators'. In an electric power-house, huge magnets and coils of insulated wire-conductors, turn around in a generator to produce electricity. Generated electric power, which is in AC form is delivered to the factories, business premises and residences, by a power distribution net-work **(Fig-2, Green line)**.

Audio, radio and video signals transmitted on electrical transmission wires are typical examples of AC electrical signals. In these applications, an important goal is the recovery of information transmitted to long distances by the AC-signals and finally to reproduce these in the original audio or video form..

13.5 Electricity and Magnetism

Electricity and magnetism are inter-related. In the year 1831, Michael Faraday, discovered the principle of dynamo. Thus, when the poles of a magnet are moved near a coil of wire (conductor), a flow of electrons takes place, and *vice-versa*, if an iron rod is held into a coil having electric current (DC) flowing through it, the rod start behaving as a magnet. This principle is used for generating electricity using a dyanamo.

Dynamos are run using mechanical energy produced by steam-engines and turbines or hydropower (flow of water under high pressure). One important advantage of AC voltage is that it may be increased or decreased, using a transformer. Transmission of AC-electricity at higher voltage leads to significantly more efficient transmission of power, with minimum loss of energy.

For transmission of high voltages efficient insulation is required and there is also increased need for safe handling (**Fig.-3)**. The magnitude of original power generated depends on the design of a generator, and it is then 'stepped up' to a higher voltage for transmission. It is again 'stepped down' to a lower voltage as desired to be used for a particular equipment.

Voltages required by the consumers, vary depending on a country and size of power-load. Generally the domestic electric motors and lighting-bulbs are built to use up to a few hundred volts (110-220) between phases.

Three-phase electrical generation is also very common. In the simplest case, it consists in having three separate coils in a generator stator that are physically offset by an angle of 120° to each other. Three current waveforms are thus produced, which are equal in magnitude. These are 120° out of phase to each other.

When coils are added, opposite to these (60° spacing), they generate the same phases with reverse polarity and so these can be simply wired together. The frequency of the electrical system varies from a country to country. Most electric power is generated at either 50 or 60 Hertz (frequency). Some countries have a mixture of 50 Hz and 60 Hz supplies, notably Japan.

As mentioned above, electricity is always generated in AC-form, at 50, or 60 cycles, and for domestic use it is made available at 110 to 220 volt in most of the countries. Since there is always some loss during transmission of electricity to longer distances, hence, such transmission is carried out at much higher (many thousand) voltage. Regularly distanced, 'step-up' and 'step-down' transformers are set up for this purpose.

13.6 Difference between Electrical and Electronic Technologies

The difference between electrical and electronic technologies is arbitrary and the scope of these two is overlaping. For engineering studies, separation of these two branches as 'electrical engineering' and 'electronic engineering' has been only four to six decades old, and mainly due to the recent explosion of electronic technologies.

Electrical engineering (technology) is primarily concerned with generation, transmission, distribution and utilization of bulk electricity. One major use of electricity is to do street lighting and lighting at public and private places. Besides this, in most of the heavy and small scale industries and transport, electricity is used *via* its reconversion into mechanical energy using electrical motors.

Numerous electric motor based machines, and other home appliances directy use electrical energy. Some of these are fans, water-pump, air conditioners, grinders and blenders, telephone, iron, hot-plates, ovens and many more. There have been many older machines and processes, which did not directly used electrical energy, e.g. sewing-machine, conventional photography, typewriter, pendulum clock, steam-locomotives, cars and other automobiles, etc. Many of these use bettary operated electronic operation and control system, or have now been modified to become based on electronics.

13.7 Electronic Technology

Production and commercialization of electricity was initially made for producing light only, but very soon it started finding applications for several other purposes. This gave rise to what we know today as 'electronic technology'.

DC-power, at variable and much lower voltage is required, for many of the electronic technologies. These sophisticated applications include most of the scientific and industrial instruements, radio and television, tele-communication, radar, sound-recording and reproduction, telephone network, computer and process-control in industry. This can be achieved, either by using batteries as the DC-source, or converting the regular AC-power supply, into DC at any required voltage, using suitable transformers and rectifiers. This has been achieved using electrical/ electronic appliances, with built-in transformers and rectifiers. Earlier, extensive use of vacuum tubes was done in electronic appliances, but to a large extent these have now been replaced by solid-state electronic devices and semi-conductors.

13.8 Role of Vacuum Tubes and Semiconductors in Electronics

Vacuum tubes have played important role in the development of electronic technology. This in turn resulted in the expansion and commercialization of radio-communication and broadcasting, television, radar, recording and reproduction of sound, analog and digital-computers, to name only a few of them. Only a few of these appliances, referred here in, had their earlier, non-electronic, i.e. mechanical counterparts, e.g. sewing-machine, pendulum clock, gramophone and typewriter. Invention of the triode-vacuum tube (and other electronic valves) and their capabilities of electron maneuvering has made many newer technologies widespread and practical. (For more details, see **Chapter-14**)

Besides the vacuum-tubes, semiconductors (transistors) have played important role in developing electronic technology. In a metallic conductor, current is carried by the flow of electrons. In semiconductors, current is often imagined as being carried, either by a flow of electrons or by a flow of positively charged 'holes' in the electronic structure of semiconductor material. However, in both these cases only electron flow is involved.

The electron flow in a semiconducting material is intermediate in magnitude to that of a metallic conductor and an insulator. This means a conductivity, approximately in the range of 10^{-3} to 10^{-8} Siemens per centimeter. Semiconductor materials are the foundation of modern electronics, which includes instruments such as the radio, computers, telephones, and many other electronic devices. More of such devices, which are based on semiconductors, include transistors, solar-cells, many kinds of diodes including the light-emitting diode, silicon controlled rectifier, and digital and analog integrated circuits. Semiconductors and solar-photovoltaic panels are devices, which directly convert light energy into electrical energy.

13.9 Conclusion

Basic differences between electrical and electronic technologies are difficult to define. In the above discussion, an attempt has been made to create a rough borderline between these. However in many other chapters of this book, which includes several electronic technologies the difference shall become clearer.

Fig.-1 Different rechargeable and non-rechargeable batteries
Author, Brianiac, Wikimedia Commons, http://en.wikipedia.org/

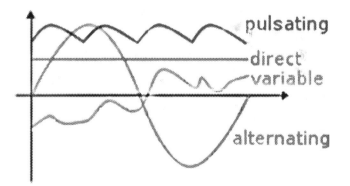

Fig.-2. Different shapes electric current
Author, Zureck, Wikimedia Commons, http://en.wikipedia.org/

P. Mathur / K. Mathur / S. Mathur

Fig.-3 High Voltage Transmission Line

Chapter-14

Vacuum Tubes, Transistors, and Solid-State Electronics

14.1 Historical, Vacuum Tubes

'Vacuum-tubes' are devices, which work by controlling the flow of electrons through a vaccum, inside a glass tube. These tubes have been used for, electric-current rectification (AC into DC), amplification, switching (on and off), for creating electronic signals (for radio and TV) and many other applications, which have been discussed in certain other chapters of this book.

All over the world, the vacuum tubes have played an important role, in the development of electronic technology. In everyday practice these are simply referred as 'tubes' or 'valves'. Earlier these were also referred as 'thermionic-valve' in U.K., and 'electron-tube' in North America.

A vacuum tube utilizes the thermionic (heat) emission of electrons from a hot metal filament or electrode, which is called a 'Cathode'. The electrons, emitted from this cathode move through the vacuum inside a sealed glass tube, towards an 'Anode' (or the positive electrode). The anode, in a tube is also referred as a 'plate'. In order to attract electrons towards it, the anode is held at a positive potential relative to the cathode. Additional electrodes (grids) are generally interposed, in between the cathode and the anode. These can alter the current (flow of electrons), giving the tube an ability to amplify and to switch, 'on and off'.

14.2 Role of Vacuum Tubes in Electronics

The vacuum tubes have played an early role in the development of electronic technology. This has resulted in the expansion and commercialization of electronic technologies such as the radio-communication, broadcasting, television, radar, recording of sound and its reproduction, spreading of world-wide telephone network, development of analog and digital-computers, and process-control in many industries.

Only a few of these applications, had their earlier, non-electronic, i.e. the completely mechanical counterparts. Typical examples of such older technologies, which did not make use of valves were, the telegraphy based on the spark-gap transmitter, gramophone, and pendulum clock. The invention of 'triode-vacuum tube' (and other electronic valves) and its capabilities of maneuvering of electron (current) flow, has made most electronic technologies widespread in use and more practical.

In most of their applications, the vacuum tubes have now been replaced by solid-state devices such as transistors, which are based on semiconductor materials. These solid-state devices are long-lasting, much smaller, more efficient and reliable, yet cheaper, when compared to equivalent vacuum tube devices. They use far less electric-power, compared to the valves. This can well be realized from the fact that the size of a domestic radio (receiver), television set, and visual display devices of many instruments, now have a much smaller size, due to introduction of solid-state technology.

Still, there are many applications of vacuum tubes, for which solid-state device substitutes have either not been developed, or are impracticable. Hence specialty vacuum tubes are still manufactured for such applications and as replacement for those, being used in certain existing equipments such as high-power radio transmitters.

Looking to a very large, domestic market for vacuum tubes in India, during the early 1950's, the then Government of India, installed a unit for exclusive manufacturing of different types of valves, at the Bharat Electronics Limited (BEL). This gave a big boost to electronic industry in India. Electronic valves of varied specifications were produced at BEL, but now it has stopped manufacturing of such valves.

14.3 Classification of Vacuum Tubes

Vacuum tubes, which have only two active components (electrodes), i.e., a filament (cathode) and a plate (anode) are referred as 'diodes' and these are mainly used for current rectification. Those vacuum tubes, which have three, ormore than three active elements are respectively called 'triodes', 'tetrodes' and 'pentodes' etc. In **Fig.-1,** a variety of typical vacuum tubes are depicted. Some of these are used for signal amplification.

Amplification of electronic signals such as those of an oscillator, and on-off switching are conveniently done using a diode.

Most common vacuum tubes in use are triodes and these have three components, i.e. a cathode, a grid, and an anode (plate). These are further classified according to their applications and their requirements. Thus, their classification can be based on frequency ranges, i.e. audio, radio, very high frequency, ultra-high frequency and microwaves. According to the power rating, they are further classified as those used for small signal, audio-power, and high-power radio transmission. Yet another classification is based on design purpose, e.g. sharp *versus* remote cutoff; amplifying, switching, control and signal amplification, rectification, mixing, general-purpose *versus* very low micro-phonic and low noise audio amplification, and so on. There are no sharp distinctions in these classifications. As an example, dual triodes can be used for audio pre-amplification and as flip-flop, in computers. Linearity is an important requirement in the former case and a long life in the latter.

In a very broad sense, there are also certain other vacuum tubes, which have quite different construction and different functions. Typical examples of these include a cathode ray tube. In this case an electron beam is produced for visual display of a cathode ray tube in a TV-set or a computer monitor.

Some vacuum tubes can also have very specialized functions, such as those in electron microscopy and electron-beam lithography.

A specialty built vacuum tube can also be X-ray tube.

Functioning of a photo-tube, and a photo multiplier tube also depends on the flow of electrons through a vacuum. Here the emission of electrons from the cathode depends on energy of a photon. Hence this is not considered to be a case of thermionic emission of electrons.

Description of such vacuum tubes, having functions other than electronic amplification and rectification, has been excluded from this article.

14.4 Construction Details of Vacuum Tubes

This has been largely described earlier. An electronic vacuum tube can have a minimum of two or more electrodes. These electrodes are fitted inside an evacuated glass tube. The glass tube is provided with a ceramic base, which acts as an insulator. A metal envelope is provided at the top.

A required number of leads, are connected to various electrodes in a tube, and these pass through the metal envelope *via* an airtight seal. Most of these tubes have leads at the bottom, in the form of pins, which can be plugged into a tube socket. Thus these tubes can be replaced, if fused or when desired.**14.5 Working Details of Vacuum Tube**

The vacuum tubes have in fact evolved from incandescent light bulb (described in **Chapter-2** of this book). Just like a bulb, a vacuum tube also contains a filament, which is sealed into a glass envelope. This filament, or the cathode, on heating, releases a stream of electrons into the vacuum, inside the tube. This process is called 'thermionic' (heat) emission of electrons. A second electrode, which is called a plate, acting as an anode, attracts these electrons, because of its positive potential, with respect to the anode.Thus there is a net result in a flow of electrons from filament (cathode) to the plate (anode). The function of filament is to emit thermionic electrons, which alongwith the plate, creates an electric field, arising due to the potential difference between these two electrodes.

The term rectification of current, means conversion of AC into DC. For rectification of AC into DC, such vacuum tube with only two electrodes or a diode, can beused. Since direct current travels unidirectionally, a diode in such case only converts AC into pulsating DC. Hence it can be used in a DC power supply. A diode is also used as a demodulator of 'amplitude modulated' (AM) radio signals and have other similar functions.

In early vacuum tubes, a directly heated filament was used as cathode, but in modern tubes indirect heating of cathode is done. In such cases a separate electrode is used as a 'hollow cathode'. Inside such hollow cathode is installed an electrically insulated, heating filament. In such arrangement, the insulated heater does not function as an electrode, but simply serves to heat the cathode, to such a temperature, at which the thermionic emission of electrons begin.Such an arrangement allows the tubes to be heated through a common circuit, which can be DC as well as AC, while allowing each cathode to arrive at a voltage independently of the others. This makes a circuit design much simpler.

14.6 Gas-Filled Tubes

There are a variety of other electric current conducting tubes, filled with a variety of gases. These gases can at a high or a low pressure. A common fluorescent tube-light and a compact fluorescent light bulb are familiar examples of gas filled tubes. These are not electronic vacuum tubes and this is not the subject of the current article.

There are yet other types of tubes, e.g. the voltage regulator tube, and thyristor-tube. Theseresemble commercial vacuum tubes. These can even fit in metal sockets designed for vacuum tubes. Their distinctive orange, red, or purple glow during their operation clearly indicates the presence of certain specific gases in the tube. The flow of electrons in a vacuum does not produce any visible light within that region. Although not properly termed vacuum tubes, they may still be referred to as electron tubes, as they do perform electronic functions.

14.7 Semiconductor Materials

Semiconductor materials, are mainly composed of such elements, as silicon and germanium. These elements are in the IV-group of Periodic Table. Another semi-conducting material is gallium arsenide. Many organic compound are also semiconductors. Semiconductors can be easily manipulated by a process called doping. This process involves the addition of trace impurities, and the process is called doping. By this process it has been possible to make useful material for electronic industry. The conductivity of these partially conducting, semiconductors can be controlled by introduction of an electric or magnetic field, by exposure to light and heat, or by mechanical deformation of a doped mono-crystalline grid. Thus, the semiconductors can make excellent sensors.

Generally, the semiconducting materials are crystalline solids, but non-crystalline and liquid semiconductors are also known. These include hydrogenated amorphous silicon and mixtures of arsenic, selenium and tellurium in variable proportions.

These compounds are better semiconductors of intermediate conductivity and their conductivity varies rapidly with temperature. Such disordered materials lack a rigid crystalline structure of conventional semiconductors such as silicon. These are generally used in thin film structures, which have far less demand for the electronic-quality of the material and thus are relatively insensitive to impurities and radiation damage. Organic semiconductors, that is, organic materials with properties resembling conventional semiconductors, are also known.

14.8 Semiconductor Devices

Transistors are devices, which are based on the electronic properties of semiconducting materials. They are used as electronic-components in radio and many other electronic instruments. Electrical conductivity of a semiconductor material is due to electron flow of a magnitude, which is intermediate magnitude between that of a normal metallic conductor and an insulator, e.g. certain plastics. Conductivity of a semiconductor is approximately in the range of 10^3 to 10^{-8} Siemens per centimeter.

The very foundation of modern electronic devices, e.g. transistor radio, liquid-crystal display based TV-sets, computer monitor, telephone, and many other devices are based on the use of semi-conductor materials. Other semi-conductor based devices include transistors, solar cells, many kinds of diodes including the light-emitting diode, silicon controlled rectifier, and digital and analog integrated circuits.

14.9 Historical, Developments in Electronics

J. E. Lilienfeld, who was a physicist, in the year 1925, filed a patent for a transistor in Canada. He described this device similar to a 'field-effect transistor' (FET), but did not give more details of his invention. Later, in the year 1934, a German inventor, O. Heil also patented a similar device. During the year 1942, H. Mataré, while working on a detector for a 'Doppler Radar' system, built and experimented with a devise, which he called 'duodiodes'. This devise had two separate, but close metal contacts on semiconductor material. He discovered certain effects that could not be explained by two independently operating diodes and thus it formed the basic idea for 'point contact transistor'.

Two inventers, namely J. Bardeen and W. Brattain, in the year 1947 were working for the Bell Telephone Laboratories of USA. They observed that when electrical contacts were made to a germanium crystal, the output electrical power was much larger than the input power. W. Shockley, who was the Group-Leader of the 'Solid-State Physics Group', conceived a big potential in this observation. More work, which he carried expanded our present day knowledge related to semiconductors. Technologist, J. R. Pierce, for the first time used the term 'transistor, which meant 'transfer resistor'.

14.10 Electron flow in Semiconductor

Semiconductor based solar photovoltaic cells are capable of directly converting light energy into electrical energy. In a simple metallic conductor, current is carried by a flow of electrons, while in case of a semiconductor, current is supposed to be carried either by the flow of electrons or by the 'flow' of the so called 'positively charged holes', which are present in the electronic structure of the material. In reality however, in both the cases only the movement of electrons is involved. Semiconductor devices have now largely replaced electronic vacuum tubes in most of the applications. Unlike the valve, the electron flow in semiconductors takes place in solid state and not through a high vacuum in glass-tubes.

14.11 Manufacture of Semiconductor Devices (Fig.-4)

Manufacturing of semiconductor devices is done both as single discrete devices, as well as an integrated circuit (IC). In the later case a single unit consists of a number of devices, which can range from as fewer as ten to several billions. Such devices are manufactured and interconnected on a single semiconductor substrate, which is called a 'chip'.

The two types of transistors have some difference in the way they are used in an electronic circuit. A 'bipolar transistor' has its terminals labeled as 'base', 'collector', and 'emitter'. A small current at the base terminal, which flows from the base to the emitter, can control or can switch a much larger current between the collector and emitter terminals. For a 'field-effect transistor', the terminals are labeled 'gate', 'source', and 'drain'. The voltage at the gate can control a current between the source and the drain.

The primary advantage of a transistor lies in its ability to use a small signal applied between one pair of its terminals, to control a much larger signal at another pair of terminals. This property is called a 'Gain-Effect'. A transistor can control its output, which is proportional to the input signal. Thus, it can act as an amplifier. Alternatively, the transistor can also be used in an electronic circuit to turn a current 'on or off" by an electronically controlled switch, where the amount of current is determined by other circuit elements.

14.12 Conduction of Electric Current in Semiconductors

Mobile electrons and the holes, referred earlier, in a semiconductor are collectively called as charge carriers. The number of free electrons or holes in any semiconductor can be increased by doping a silicon based semiconductor with a small amount of impurity atoms, e.g. those of phosphorus or boron.

There are two types of semiconductors, which are called the 'p-type' and 'n-type'. A semiconductor is used for a doped semiconductor, which contains an excess of holes, while those, which contains an excess of free electrons is called 'n-type'. Here the prefix 'p' and 'n' stands for positive holes and for electrons (negative), respectively. These signs indicate the charge of the majority mobile charge carriers. The semiconductor material used in a device is doped under highly controlled conditions at their fabrication facility, to precisely control the location and concentration of p—and n-type doping agents. A junction, which is formed at a site, where n-type and p-type semiconductors join together is called a p-n junction.

A transistor is a typical semiconductor device, which is used for amplification of current and a switching device (on and off) for electronic signals. It is made up of a semiconducting material with three or more terminals, by which it can be connected to any external circuit. A voltage or current, which is applied to one pair of the transistor's terminals changes the current flowing through another pair of terminals. Since the controlled, power output can be much more than the controlling, or the input-power, a transistor can act as an amplifier for the input signal. Today, some transistors are packaged individually, but most of them are kept embedded in integrated electronic circuits.

14.13 Conclusion

To sum up therefore, transistors are the fundamental building block of all modern electronic devices and systems. Commercial introduction of transistors in the beginning of year 1950 and onwards, the transistors have revolutionized the field of electronic engineering. This has resulted in manufacturing much of smaller and cheaper electronic equipments, such as radios electronic calculators and, and computers.

Fig.-1, Different types of vacuum tubes
Author, Stefan Riepi, Wikimedia Commons, http://en.wikipedia.org/

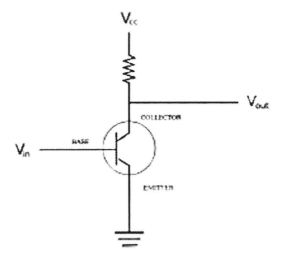

Fig.-4, A simple circuit diagram of a bipolar transistor.
Author, Stefan Riepi, Wikimedia Commons, http://en.wikipedia.org/

Chapter-15

Microwave Oven and Electric Oven

15.1 Introduction: Man Learns Cooking

Just like his fellow animals, primitive men ate raw food, which he found in the nature surrounding him. But unlike most animals, he has already developed strong taste buds, which made him very selective in choosing his food, based on its flavor and taste. Man also became an early gatherer of food, which he started storing, after his settlement small community groups.

Due to human curiosity, or perhaps, while sitting by the fire, which he always kept near his habitat, he might have put some food (pieces of meat or food-grains) on burning charcoal. To his surprise, he might have discovered a likable change in the flavor and texture of that food item. Food may have become more flavored, better tasting, likable and easily digestible for him. Then onward he deliberately started treating his food by heating on fire. Such might have been the beginning of a 'new technology', now known to us as 'cooking'.

Except the human being, no other animal does any pretreatment, or cooking of his food, such as grinding, blending of food constituents, and cooking by heat. Over centuries, human digestive system has so adopted to the use of cooked food, that he can no longer eat raw food, except perhaps some well-ripened fruits. Man has used cooking technology, not only for his own food, but also for the food, he gives to his domesticated animals (horses, cattle, dogs, cats and birds). Today cooking of food and marketing of cooked and packaged food has emerged as a very big industry and technology all over the world.

15.2 Kitchen and Cooking: Women's Domain

During all the times and among all the human races, there has always been a division of work between men and women, though such division of work is fast disappearing now. Cooking has mainly been in the domain of women's work. With advancement of civilization, the technology of cooking has also progressed very well, and cooking has now become a profession. This has also necessitated several improved cooking facilities in a kitchen. Of these, the most important kitchen facility has been a 'cooking oven' (Hindi, *Chula* or *Tandoor*), which has conventionally utilized burning of firewood and charcoal for heating.

15.3 Conventional Ovens

An oven is used for cooking, which involves heating or backing of food. Generally, an oven is a thermally insulated chamber, which is installed in a kitchen. Ovens have been in use, ever since prehistoric times and by all different civilizations. The design of an oven has been a novelty of each human race and culture. Hardened mud ovens were generally used all over the ancient world.

Many ancient civilizations, particularly the Indian and the Greeks have been credited to develop the process for baking of bread, into an art. While the Greeks created a wide variety of dough and loaf shaped breads, round and flattened breads (*Roti* or *Chapati* and *Paratha*), using a variety of powdered edible grains and their mixture were used for making '*Roti*' in India. '*Tandoor*' has also been a typical oven, used for backing in India.

15.4 Electric oven

After the invention of electricity and establishment electric power generation and distribution system in cities and towns in the western countries, a limited use of electric energy (*via* heat) for cooking began. One of the electrically heated cooking appliance, which man developed was 'electric oven'.

Electric ovens **(Fig.-1)** have also been important equipment in scientific laboratories. These ovens, though in a slightly modified form became a common facility in kitchens. Such ovens can generally be heated to a temperature of 150-300°C, using electric power, *via* heating coils, which are provided in them. Generally it is possible to control the temperature inside an oven to within plus or minus 5°C, by a thermostatic control mechanism.

In a kitchen, an oven is used for cooking certain recipes, e.g. cake, pizza and many Indian recipes and reheating of earlier cooked food. Like any electric heating device electric power consumption of an oven is quite high. As a kitchen facility, these ovens are now being phased out and being replaced by power saving 'Microwave Ovens', which also have many other advantages over conventional electric ovens.

15.5 Working of a Conventional Electric Oven

In an electric oven, the current carrying heating coils transfer heat (by conduction, convection and radiations) to the enclosed air, which can ultimately acquire a temperature, desired for cooking a particular recipe. Cooking is then achieved by keeping the food, which is placed in cooking trays, vessels, or directly on the shelves in an oven, and heated to a desired temperature, for a prescribed time.

15.6 Microwave Region of Electromagnetic Spectrum

Microwaves are a part of the electromagnetic spectrum. The total region of electromagnetic radiations extends from extremely high energy cosmic and X-rays, through the ultra-violet and visible portion of the spectrum and finally to the infra-red, microwave and radio-frequency region.

The energy of the radiations in any particular region of the spectrum is given by the relation—E=hv, where, v= frequency of the radiations = 1/λ, where λ is the wave-length and h is Plank's constant.

This means that, smaller the wave-length of radiations, higher is the energy, The region of microwave radiations of the electromagnetic spectrum lies, in between the common radio-waves and infrared radiations. Of these radiations, the microwave-radiations, (but not the radio-waves, have sufficient energy for excitation of the translational and some vibrational modes of motion in polar-molecules. However, the microwave portion of electromagnetic radiation, do not carry that high energy per quantum, which is required for electronic-excitation or for ionization of atoms or molecules.

15.7 Discovery of the Heating Effect of Microwaves

Like many of the today's great discoveries, leading to the development new technologies, the use of microwave for heating (cooking) was invented accidentally. This was during a radar related research project in the year 1946. While working on a completely different technology, Dr. Percy Spencer **(Fig.-2)**, who was a self-taught engineer, noticed something very unusual during his experimentations.

At that time, Dr. Spencer was working for the Raytheon Corporation (USA). He was testing a new source of producing radar-waves, which was called 'magnetron'. To his surprise, he observed that a candy bar, which was present in his shirt pocket, has melted during the experimentations.

Dr. Spencer was very much intrigued by this observation, so he tried yet another experiment with magnetron. This time he placed some popcorn kernels near the magnetron, while he stood a little farther away from it. Again to his surprise, he observed, with an inventive sparkle in his eyes, that the popcorn sputtered, cracked and popped up, all over his lab.

The vary next morning, Dr. Spencer, along with another curious scientist colleague, whom he has invited to his lab, decided to put the magnetron tube near an egg. This time, Dr. Spencer along with his scientist colleague, observed that the egg began to tremor and quake. Both of them observed a very rapid rise in the temperature of egg-yolk, inside the shell, which was causing tremendous internal pressure. It was just at that moment that the curious colleague of Dr. Spencer moved nearer, for a close look of the egg. Then suddenly the egg exploded, shattering, hot egg-yolk all over the laboratory and on the amazed faces of the two scientists.

On looking at such a mess, which has been created in the laboratory, the face of Dr. Spencer lit up with a logical and scientific conclusion. He concluded that, melting of a candy bar in his shirt pocket, popping up of corns, and exploding of an egg-shell, were all due to heating effect, which was caused by exposure to low-density microwave energy, produced by the magnetron.

Thus concluded, Dr. Spencer-

"When an egg could be cooked so quickly, on exposer to microwaves; there should also be a possibility of using these radiations for general cooking".

And thus, a series of experimentation on heating effect of microwaves began in Dr. Spencer's lab.

15.8 Dr. Spencer's Microwave Oven

Next, Dr. Spencer designed a metal box into an oven, which had an opening to feed microwave energy into it. The microwave energy, which was fad into the oven, remained confined into the box, and it created a high density of electromagnetic field, within the box. When a tray of food was placed inside the box, and microwave energy was fed into it, the temperature of the food rose more rapidly, when compared to that of the enclosed air. Dr. Spencer had, thus invented a 'microwave oven' **(Fig.-3)**, which was to revolutionize cooking, and which, now forms the basis of today's multimillion-dollar industry, manufacturing of microwave ovens.

15.9 Principle of Microwave Oven

A microwave oven is based on 'dielectric-heating' for cooking of the food. It has been well known that microwave radiations preferentially interact with polar molecules. These polar molecules can be those of water and the molecule of carbohydrate and proteins in the food. Molecules of these essential constituents of food contain a large number of polar groups, e.g., hydroxyl and amino groups ($-O^{-\delta} H^{+\delta}$, and$-N^{-\delta} H^{+\delta}$) in their molecules. Molecules, containing such polar groups, have a strong property of absorbing energy in the form of microwaves, which cause an increase in their rotational, vibrational and translational energy. These are the energy form, which constitute heat.

Transfer of heat by conduction through the walls of a cooking vessel or by convection through the fluid content in a food is normal in a cooking process. Unlike such heat transfer, excitation of polar molecule, in food material by microwaves is direct, i.e. at the molecular level. Such molecular excitation this is fairly uniform, throughout the bulk of food. This, leads to more uniform heating, when compared to the conventional cooking methods, where heat transfer takes place, by conduction through the walls of a metallic container and into layer after layer of food.

Microwave heating is more effective on liquid water (in food), than water in frozen foods, where the molecules are not so free to rotate. In case of proteins, fats and carbohydrates, which have a small molecular dipole moment; compare to water, microwaves are slow to interact.

Sometimes, the microwave heating is explained as a resonance of water molecules. This is scientifically is incorrect. Resonance only occurs in water vapor, and at much higher frequencies, i.e., at about 20 GHz. All domestic as well as large industrial and commercial microwave ovens, operate at a common microwave heating frequency of 915 MHz, (wavelength 328 millimeters), which heats water and food perfectly well.

15.10 Induction Heating

Here, we shall make a mention to another electric heating device, which has now been adopted for cooking. It is called 'induction heating'. Earlier, some hot plates having encased and insulated electrical heating coils were used, in science laboratories as well as in kitchens. Since the heat transfer by a conventional hot plate takes place *via* heat conduction from the plate surface, there is a simultaneous loss of heat by radiations from the surface of the hot plate.

To avoid such loss, metal foundries have long been using induction heating to melt metals employing 'induction furnaces'. By induction heating, heat is directly transferred into the metal. More recently, hot plates have been designed based on direct, 'induction heat transfer' into the food. With these specialty designed hot plates, the electrical energy is directly transferred into the food, for cooking, thus economizing power consumption.

15.11 Microwave Cooking

One limitation of microwave heating is that it can cause localized thermal runaways in some materials which have low thermal conductivity. Dielectric constant increases with temperature and under certain conditions, glass can exhibit thermal runaway in a microwave even to the point of melting. Hence, glass-wares are not preferred for microwave cooking.

There is also a common misconception, that microwave ovens cook food, 'from the inside to out', which means from the center of the entire mass of food to outwards. In reality, microwaves are absorbed in the outer layers of food in a manner somewhat similar to normal heat transfer by other methods, e.g. conduction. This misconception arises from the fact that microwaves can penetrate, through dry and non-conductive substances, which are generally present at the surfaces of some common foods. Thus microwaves often induce initial heat 'more deeply', when compared to other methods of heating, for cooking. Depending on the water content in a particular food, the depth of deposition of heat (energy) into it can be several centimeters more in a microwave oven, compared to normal cooking.

Thus, there is a difference between simple broiling by infrared radiations, which are produced by burning of charcoal, where layer by layer transfer of heat from the surface of food takes place, and microwave cooking. Penetration depth of microwaves is dependent on the composition of a particular food item and the frequency of microwaves. With lower microwave frequencies of longer wavelengths, the penetration is much deeper. In a different sense, microwaves 'do cook' from 'inside out' because each molecule in food is generating heat from 'inside' (molecule) and radiating it 'outward', towards the surrounding food (out-side).

15.12 Wave-length and Heat-Effect

A microwave oven generally uses, microwave radiations of frequency of 2.45 gigahertz, (GHz, giga = one billion), which corresponds to a wavelength of 122 millimeters. Water, fat, carbohydrates, proteins and many other constituents of the food absorb energy from microwaves in a process, which is called 'dielectric heating'. Many molecules in food, e.g. those of water are electric dipole. It follows that they have a fractional positive and negative charges at the two ends of a molecular-dipole. Such molecular species, start rotating as they tend to align themselves with the alternating electric field of the microwaves. This molecular motion, which constitutes heat, is then dispersed as the rotating molecules hit each other, causing translational and vibrational motions.

15.13 Start of Microwave Oven era, and its Earlier Limitations

Though the first use of microwave for home cooking application was introduced by Tappan Company (USA) in the year 1955, the manufacturing of microwave ovens, for personal use was started much later, i.e. in the year 1967, by the Amana Corporation in USA.

Microwave ovens heat food, quickly and efficiently, but, unlike the conventional electric ovens, they do not bake the food to the stage of browning it. It makes these ovens, somewhat unsuitable for cooking certain foods, or to achieve certain culinary effects of conventional cooking. Hence, certain additional type of heat sources are frequently installed into microwave ovens, to get these, additional culinary effects.

15.14 Early era of Microwave ovens

On October eighth, of the year 1945 Raytheon Co. filed a US patent for Spencer's microwave cooking process. This was followed, in the year 1947, by placing a Raytheon Microwave Oven in a restaurant, in Boston City for initial testing. Raytheon Company also built another microwave oven and named it 'Radarange Oven'. This was the first commercial microwave oven in the world. The oven, which weighed nearly 340 kg was 1.8 meter high. The cost of the oven was about $5000 (US) per piece, with a power consumption of 3.0 Kilowatts. In contrast to this, a modern microwave oven is much smaller in its size and uses only about one-third as much power.

A Radarange Oven was also installed, in a galley of a nuclear-powered, US passenger-cum-cargo ship, the 'NS Savannah'. It is still said to be installed there, as an antique show-piece. Another commercial model of microwave oven was introduced in the year 1954, which had a power rating of 1.6 kilowatts and its cost was $2000 to $3000.

Raytheon Company, licensed its microwave-technology to the 'Tappan Stove' Company of Mansfield (Ohio, USA), in the year 1952. Mansfield Co. tried to market a large, 220 volt, home microwave oven in the year 1955 for a price of $1295, which did not sell well. In the year 1965 Raytheon Co. acquired Amana Co. and in the year 1967, they introduced the first popular, domestic model of a microwave oven, which outdid earlier model, Radarange. The price of a microwave oven had now come down, close to $500.a piece.

There has been a general practice in USA and Canada to furnish the kitchen in a rental apartment with a cooking-range, an oven and a refrigerator, but till mid-1980s microwave ovens were not a common feature of kitchen furnishing, in rental apartments in USA. These were generally owned by the person, who rented an apartment. Kitchens of some rental apartments in Indian metros are now furnished with microwave ovens.

15.15 Some General Features of Modern Microwave Ovens

Modern microwave ovens **(Fig.3)** are provided with a control panel, with LED (light emitting diodes), which is based on liquid crystal, or vacuum fluorescent display technology. The control panel keypad always contains 'Start and Stop' buttons. These, numeric buttons are meant for entering the cooking time, and for selecting the power levels, which can be high, medium, or low, and a defrost button. Additional buttons may be present, to choose the type of food to be cooked, i.e., meat, fish, poultry, vegetables, frozen foods, and popcorn,. Some high-priced models additionally have a 'sensor cook' button as well. The display panel can generally show the time of a day, adjustment of which varies from model to model. This is generally necessary after a loss of power or for seasonal time changes.

15.16 Expending Market of Microwave Ovens

Litton Co., in the year 1960, purchased Studebaker's Franklin Manufacturing Co., which had been manufacturing magnetrons and microwave ovens. Litton then developed a new configuration of the microwave beam, which was short and wide in shape. The new oven helped in a rapid growth of the market for home microwave ovens; the sales reaching to nearly one million pieces per year by the year 1975.

The microwave source, i.e. the magnetron, was also redesigned in Japan, which allowed manufacturing of less expensive oven units. Soon the manufacturing of microwave ovens was also started in India, where it has now become an affordable kitchen accessory, and microwave ovens are now seen as a common feature in kitchens in metros and towns in India.

Fig.—1. A modern electric oven
Author, Arnold Reinhold, Article Cooking ovens,
http://en.wikipedia.org/

Fig.-2 Dr. Percy Spencer, Discoverer of Microwaves
Article Microwave ovens, http://en.wikipedia.org/

Fig.-3. A Typical Microwave Oven
Author, Doris R Hanin, Article Microwave ovens,
http://en.wikipedia.org/

Chapter-16

❖

Visual Aids (Old and New) for Teaching and Lecturing

16.1 Introduction

Visual Aids can make learning experience extremely interesting and memorable, if used during lecturing and teaching. This, in fact is true for a teacher as well as for a learner. Teaching aids are things, which are to be used in a classroom to improve teaching as well as learning. Visual aids have long been used in most of the modern educational institutions and these are the most important facilities, required in good schools and colleges.

The visual aids fall in two main categories, i.e.-

1. Those, which are used for overhead projections and;
2. Interactive tools such as a video programs and resource-packs, which are frequently distributed among the learners.

There is no doubt that a good teacher can very well describe his subject verbally, and even without any illustrations, but visual aids make his task much more effective and easy. This is particularly true for teaching of sciences, medicine, engineering, mathematics and geography like subjects. Use of regional map-charts has been a very well-established technique for teaching geography, while extensive use of roll-up charts, illustrating various processes and instruments has extensively been made in teaching basic sciences i.e., chemistry, physics and life-sciences.

Sometimes, adoption of too many different themes in teaching with the help of visual aids can confuse students in a class, and it is better to stick and follow fewer techniques.

16.2 Blackboard: The Traditional Visual Teaching Aid

Blackboard and chalk-sticks (including the colored ones) have been the most traditional teaching aids. These are being used from elementary schools, right up to college and university level. A teacher, teaching a particular topic during each session and to each class uses these facilities repeatedly, and most frequently. Blackboard, in a classroom is used for writing as well as for illustration by diagrams. Hence blackboard has been one of the most 'traditional visual aid' in teaching.

The capabilities of teachers generally differ in the use of a blackboard, and at times it can also be time consuming, to use a black board. A teacher's own handwriting and drawing capabilities differ from person to person, and it is very much reflected in his use of a blackboard.

Besides the blackboard and roll-up charts and certain more modern visual aids, e.g. overhead projections are now becoming increasingly available for classroom teaching.

16.3 Screen Projected Visual Teaching Aids

Visual aids, described here are those, which support presentations in the form of texts, tables, cartoons, graphs, illustrations and photographs. These can be the transparencies used for overhead projections, pre-printed handouts, roll-up charts, posters, and many others. These offer great help in breaking up the monotony of a lecture class and provide a visual stimulant to reinforce, which a learner needs in addition to hearing.

Projection of any earlier prepared material, on a screen, has been the most commonly used visual aid. There are four main techniques for projection onto a large screen, so that everyone in a class can see it. A presenter has a choice to select his visual aid technology, which meets the expectations of his audience. These can be-

1. Epidiascope,
2. Overhead Projector,
3. Slide Projector and,
4. Computer aided Slide Projector.

16.4 Epidiascope

No particular inventor has ever claimed the credit for designing an epidiascope. A very old, 'opaque projecting machine' has been well preserved, even now, in a lecture hall of the Cambridge University. This machine has been reported to be in use, till late 1800's. The device, which is an equivalent of epidiascope, only had an ability to project opaque objects. More modern epidiascope **(Fig.-1)**, which was developed later in early twentieth century, could project opaque as well as transparent images. Many old educational institutions in India and abroad are in proud possession of these old projectors, though the high intensity; replacement bulbs (halogen lamps, for details, see **Chapter-2**) for these may not be available in some cases.

As mentioned earlier, an epidiascope (or episcope) is an 'opaque projector', i.e. a device to display on a screen, opaque material, such as a text or a diagram from a book, a drawings on a paper sheet, mineral samples, preserved insects and plant leaves and flowers etc. Basically it has been used as an enlargement tool, which permits images of rather smaller objects, to be projected onto a screen, during a lecture or a discourse.

16.5 Working of an Epidiascope

An epidiascope makes use of a very bright light source, shining from above to produce an image of an opaque object, in a mirror, placed inside it. This is followed by a system of reflecting mirrors, prisms and lenses to focus the image of that object, onto an out-side screen. Much brighter bulb (halogen Lamp) is needed to project the reflected light from opaque objects and much larger lenses are required, compared to any other similarly used devise, e.g. an overhead projector. In order to dissipate the heat generated from a large bulb, windows and powerful fan is provided near the lamp housing. Epidiascope as an aid for visual display has largely been phased out now, in favor of other visual aids and particularly due to the development of computer added visual display facilities.

16.6 Overhead Projector

Just as in case of epidiascope, no specific person has been credited for designing the first overhead projector, but the basic idea for its design comes from epidiascope itself. Unlike epidiascope, an overhead projector is used for projecting writings and diagrams, made on transparent sheets such as glass plates, transparent plastic or cellulose acetate sheets, which are specially made for this purpose.

An overhead projector is a robust and resilient form of visual aid. This, projection technology is also cheaper and less prone to any break down. Any speaker can prepare his own transparencies, according to his choice and need. Text and figures from books can be photocopied on cellulose acetate sheets and these serve as very good transparencies, for projection. The order of projection can be changed as desired and a transparency can be dropped or repeated during a presentation according to the need of the speaker.

16.7 Construction Details and Working of an Overhead Projector

In an overhead projector, the base consists of a large metal box with windows, which contains a very bright, halogen lamp and a fan to cool it. On the top of box is a large 'Fresnel-Lens', which collimates, i.e. it renders parallel light, coming through a transparency. To the side of the box is attached a long, vertical stand, in the form of a cylindrical rod, supporting a mirror and a lens, which focuses and redirects the light onto a screen **(Fig.-2)**.

With an overhead projection, a presentation can look very professional like. It allows the speaker to manipulate it in such a way that it keeps very well, the attention of the audience. In overhead projection, there is a higher degree of flexibility, which allows a presenter to design his own text as well as pictorial illustrations.

When a transparency is placed on the top of Fresnel-Lens, the light from the lamp below it travels vertically up, through the transparency and into the mirror. From there it turns at right angle, to be focused and shone onto a screen for display. The mirror allows both the presenter and the audience to see the image at the same time. While the presenter looks down at the transparency, the audiences look forward at the screen. Presenter can use a pencil as a pointer on the transparency, which shall be shown on the screen. The height of the mirror can be adjusted, to focus the image and make it larger or smaller depending on how close the projector is kept to the screen.

16.8 Photographic-Slide Projector (Fig.-3)

The use of a photographic-slide projector has been one of the oldest, 'projection on screen' systems. Photographic-slide presentations are particularly suitable for specialized subject areas, which involves preservation and repeated reuse of the material (slides). Photographic companies e.g. Eastman Kodak Co. (USA) and Agfa Co. (Germany), make 35mm. bi-positive (color as well as black and white) films, which can be mounted as slides (transparencies). Depending on the subject matter of a presentation, it may be appropriate to use such slides.

Special equipment, i.e. a slide-projector is needed to project photographic slides. The projector uses a carousel, which operates by a trigger mechanism, and allows moving to the next slide. When the first slide has been projected for an appropriate length of time, the other slides can be projected in a sequence either manually by an operator or the presenter, with a remote control.

16.9 Computer Added Slide-Projection

Computer added slide projection, can be regarded as an extension of 'photographic-slide projection'. It is the most modern and high-tech version of slide projection, which uses a data-projector, using computer. The presentation is based on software such as 'Microsoft PowerPoint', available on a common PC.

As with any other visual aid, which relies on a specific technology, it is necessary to make sure that the equipments (computer and projector) being used, are working properly. It is desirable to pre-scan the slides and incorporate them into a 'PowerPoint' or similar presentation.

16.10 Slide Presentation Software

Most computers (PC's) provide presentation software, which allows any presenter to design a format, draft a text, and include desired photographs and illustrations for presentation. These are magnified and projected onto a screen, during a presentation. The equipment required is a desktop or a laptop computer linked to a data projector.

Computer software can also generate a variety of documents that can be used as handouts, during or after the lecture. In developing computer-based presentations there is a choice to select from a broad range of backgrounds, i.e., fonts, styles and formats. One big advantage of computer-based presentations over other techniques is that you can change the presentation very easily and there is no need to change a hard copy unless you are providing the software-generated handouts.

16.11 Conclusion

Besides the routine classroom teaching, an extensive use of projection-based visual aids is now being made in conferences and seminars, which have become more frequent since last few decades of twentieth century. How the changes have taken place in lecture projections is illustrated by the following example?

At one of the annual sessions of Indian Science Congress, in mid-twenties, Dr. Homi Bhaba (then Chief of Indian Atomic Energy Agency) gave a lecture, where lots of mathematical equations and their derivations was involved. For this he had to prepare a number of roll-up charts (canvas based black-board), on which equations were hand-written with chalk. One by one these charts were displayed before the audience, mentioning equation-number, along with chart number. Such a situation is difficult to visualize today. With the developments of computer added projection facilities and availability of so many and alternate projection of visual aids, lecture presentations are very much simplified now.

Fig.-1. A modern Epidiascope for projecting opaque objects.
Author, Berthold Werner, Article on Epidiascope,
http://en.wikipedia.org/

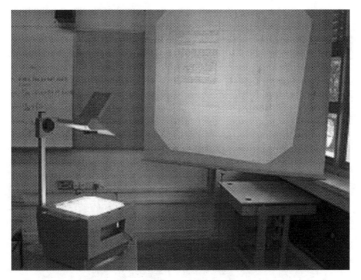

Fig.-2. An overhead projector for use in a classroom.
Author, mailer_diabio,. http://en.wikipedia.org/

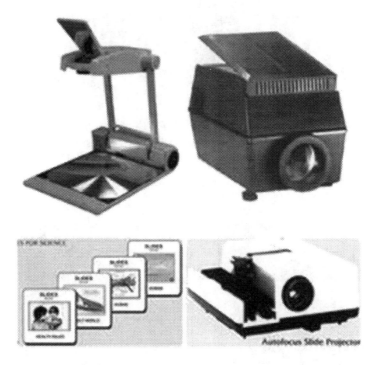

Fig.-3. An Overhead Projector, Slide Projector and Slides.
Author, Wikipedia article Slide projector and slides Wikimedia Commons. http://en.wikipedia.org/

Chapter-17

❖

Mechanical and Electronic Calculator

17.1 Historical

Much before any mechanical or electronic machine (calculator) was devised to carry-out mathematical operations, e.g. multiplications and divisions, extensive arithmetical tables were taught in rural schools in India. These were regularly recited by the students in village, elementary schools (*Pathshala* or *Posal*), usually run by a single teacher (*Guruji* or *Guransa*) at his home. Even a fresh entrant to the school was made to stand up in a line with other, senior students and recite the tables. These tables included, beside simple multiplication tables, also the multiplications involving 0.25 (quarter or *pawa*), 0.5 (half or *adha*), 0.75 (three quarter or *puna*), 1.25 (*sawai*), 1.5 (*dedha*), 2.5 (*dhaya*), 3.5 (*Ghunta*), 4.5 (*Dhancha*) and square, of whole numbers, generally up to forty. When grownup, these boys used to act like a 'living ready-reckner' or as a 'human calculator', which was the skill required to be a successful village trader.

17.2 Abacus and Early Mechanical Calculator

The oldest known tool, used for arithmetic calculations was the 'Abacus', which was devised by the Sumerians and the Egyptians around 2000 BC. Abacus **(Fig.-1)** used to be a mechanical device, consisting of a wooden frame having vertical or horizontal parallel bars on which were stung movable counters.

Much later, i.e. around the beginning of the 17th century, more tools for calculation were invented. These include the 'Geometric-Military Compass' by Galileo, 'Logarithms' by Napier, and 'Slide-Rule' by Edmund Gunterwere.

In the year 1623 Wilhelm Schickard, started the designing of a 'Pedometer', which his professional craftsman could never finish. The year 1642 saw the invention of first mechanical calculator, by a Frenchman, Blaise Pascal, which came to be known as the 'Pascal's Calculator'.

In the year 1960 an incomplete replica of a mechanical calculator, based on Schickard's drawings was prepared. Even with many improvements during the 20th century, it was found inefficient. Schickard's design, with a single-tooth gear was not an adequate mechanism for a mechanical calculator. However the mechanical calculators, developed during the 17th century, continued to be refined into more sophesiticated machines, which ultimately took the shape of the analog computers.

17.3 Mechanical Calculating Machines

Three hundred years ago, Pascal, when he was only nineteen years old, invented a calculating machine. This was necessitated due to the fact that he was very much bothered to see the burden of arithmetical work, which his father had to do. His father was working as a supervisor of taxes in a office at Rouen. Pascal conceived the idea of doing the work mechanically, and he developed and designed an appropriate machine for this purpose.

Pascal's invention of calculator at that time was too premature. During those days the technology of mechanical engineering was not much developed and technology existing during those days was not sufficiently advanced to enable his machine to be made economically. Accuracy and strength, needed for a long time use of Pascal's-Machine could not be achieved. To conceive an idea was one thing, but to actually design the machine, and manufacture it commercially was difficult. It was not until the nineteenth century, that a renewed stimulus to Pascal's invention was given and a workable mechanical calculator was developed.

17.4 Slide-rule: A Simple Calculating Device

A slide-rule **(Fig.-2)**, invented by Edumond Gunter is a simple device consisting of two or more logarithmically scaled rulers, which are so mounted as to slide along each other for carrying out many mathematical operations, such as multiplication and division. With the help of a slide-rule, more complex computations could be reduced to addition and subtraction. Even today, handy slide-rules are extensively used on site, by the technocrats, when other office facilities are not available to them. Besides being a calculator, a slide-rule functioned as a substitute for the book-let of mathematical tables, which included tables for logarithm, and trigonometric-functions.

Till mid-1960's, in the science, engineering and other technical institutions, the students were encouraged to use slide-rule and some training on its use was also imparted as a part of curriculum. With an extensive introduction of electronic calculator after 1970's, slide rules gradually gave way to the electronic calculators.

17.5 Electronic Calculator

A simple electronic calculator is a device, which can perform the basic mathematical operations i.e. addition, subtraction, multiplication and division. Compared to a table-model 'personal computer', a pocket size, hand-held electronic calculator is easy to carry to any place. A scientific electronic calculator **(Fig.-3)**, besides being is much smaller in size, is also very easy to use. Modern electronic calculators can vary from credit-card sized cheap models to sturdy desktop models, with a built-in printer, to keep a record.

Pocket-sized calculators became commercially available in 1960s, when microprocessors were developed, by the 'Intel' and used for business and commercial establishments. These calculators became even more popular during mid 1970s, when integrated circuits made them much smaller and cheaper. By the end of 1970's, the prices of calculator were reduced to an extent that a basic calculator became affordable even to small traders and to the students.

Besides general purpose calculators, there are also calculators for specific uses. Thus; there are the scientific calculators which can perform calculations based on logarithms, trigonometric functions and statistical functions. These calculators can have an ability to do computer algebra, while the graphic calculators can be used to graph functions defined on the real line, or higher dimensional Euclidean-Space.

17.6 Use of Calculators in Educational Institution

By the year 1986, calculators represented an estimated 41% of total general-purpose office hardware for commercial computation. By the end of the year 2007, and with the introduction of PC's, this has now again been reduced to an insignificant number. Yet, in most of the countries, science and engineering students still use calculators for their class-room mathematical work.

Since a high degree of mathematical proficiency is considered necessary among the science and engineering students, initially there was some opposition to the idea of permitting calculators in educational institutions in India. It was argued that the use of calculators shall hinder the development of basic skills of mathematics among the students. Inadequate guidance in the use of calculator can further reduce the mathematical thinking, which is required by the students. Some educationists have argued that exclusive calculator use can reduce mathematical skills and prevent understanding of advanced algebraic concepts. Even today, there is a difference of opinion, among the educationists about the importance of the ability to perform calculations in once brain and without the help of a calculator.

17.7 Electronic Calculator: Components and Working

A modern electronic calculator consists of the following main components:

1. A power source (battery, it can also be a solar cell.
2. Keypads, which consists of keys used to input numbers and function commands (+,—, x, ÷, log, sin, etc.).
3. Display panel, where the input numbers, commands and the final results are displayed,. Seven stripes (segments) are used to represent each digit in a basic calculator.
4. Scanning unit, consisting of a microprocessor chip. When a calculator is in use, it scans the keypad which picks up an electrical signal when keys are pressed.
5. Encoder unit, which converts the numbers and functions into the corresponding binary code.
6. X-register and Y-registers, which store numbers where these are temporarily stored, while doing calculations. All the numbers first go into the X-register. The numbers in the X-register are shown on the display.
7. The functions for any calculation are stored, until the calculator needs these for flag register..
8. The instructions for in-built functions, i.e. arithmetical operations, square roots, percentages, trigonometry etc. is stored in a binary form of permanent memory. These instructions are the 'programs' stored permanently and cannot be erased.
9. User memory (RAM) is the store, where numbers can be stored by the operator; and these can be changed, as desired by the operator.

10. The Arithmetic-Logic Unit (ALU), executes all arithmetic and logic instructions, and provides the results in a binary coded form.
11. Finally a decoder unit converts the binary code into 'decimal numbers' which can be displayed on the display-panel.

17.8 Manufacturers of Electronic Calculators

During the years 1960-70s electronic calculators have undergone very rapid developments. Initially light emitting diode (LED) display was used, which consumed more power resulting in a short battery life. LEDs were replaced by liquid crystal display during early 1970s. In the year 1973. Texas Instruments Company introduced their SR-10 model electronic calculator, which was also called 'electronic slide rule', during that time.

The cost of a scientific pocket calculator at that time was about $150. Shortly more SR series models were introduced, while the first programmable pocket calculator (HP-85) was introduced by Helwet and Pacard in the year 1974. It had a capacity of 100 instructions, and could store and retrieve programs with a built-in magnetic card reader.

17.9 Conclusion

Educationists may still differ and plead, 'for or against', the use of electronic calculators, but these electronic facilities have now evolved, and these are here to stay, and to be used. Most educationists strongly favor the development of at least simple mathematical skills in students and even a common man. This is true and necessary. Others may argue, 'If a simple machine can do such simple mathematics why bother your head with it?'

Why bother? Because a child's physical and mental capabilities are not fully developed, when he is born. Some mathematical skills are necessary for mental development, e.g. to know that, two plus two is four and not five.

Fig.-1 A simple abacus used in schools

Photo Neo, Source Danish School, Article on Abacus., Wikimedia Commons. http://en.wikipedia.org/

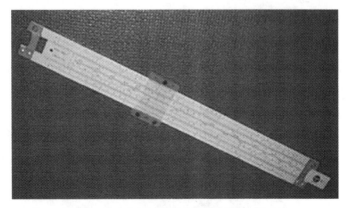

Fig.-2 A typical student slide rule

Author A Reinhold, Article on Slide Rules, Wikimedia Commons. http://en.wikipedia.org/

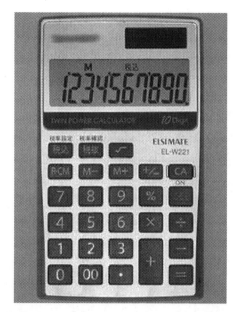

Fig.-3 Battery operated Electronic pocket calculator
Author Spring Days, Article on Electronic Calculators,
Wikimedia Commons. http://en.wikipedia.org/